Big Black Boring Rock

Essays on Northwest Geology

by

Stephen P. Reidel

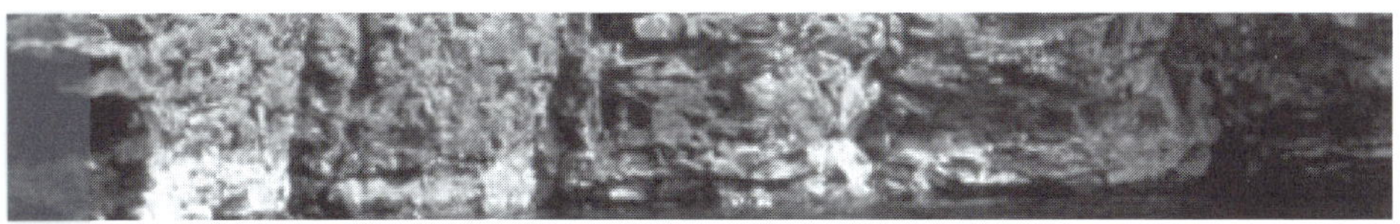

Library of Congress Cataloging-in-Publication Data

Reidel, Stephen P.
Big black boring rock : essays on Northwest geology / by Stephen P. Reidel
p. cm.
Includes bibliographical references and index.
ISBN 1-57477-156-6 (pbk. : alk. paper)
1. Geology–Northwest, Pacific. I. Title

QE79.R45 2006
557.95–dc22

2005057064

Cover Design: Jane F. Winslow, WinSome Design, Inc.
Book Design: Jane F. Winslow, WinSome Design, Inc.
Illustration: WinSome Design, Inc.
Author Photo: Rachel Berkowitz

Printed in the United States of America

505 King Avenue
Columbus, OH 43201-2693, USA
(614) 424-6393; (800) 451-3543
Fax: (614) 424-3819
http://www.battelle.org/bookstore
Email: press@battelle.org

Acknowledgments

I would like to thank the *Tri-City Herald* for providing the opportunity to write the Northwest Geology column from which the essays in this book were developed. The support of Don McManman, Chris Sivula, and Rick Larson over the many years is greatly appreciated. I also would like to thank Dr. Duane Horton and Mimi Knight who read drafts of those columns and provided invaluable reviews. Their reviews made the science readable. Thanks to Dennis Dauble for supporting this project. And finally, I owe a great thanks to Georganne O'Connor and Jane Winslow who did all the work to design, edit, and assemble this book so that I could get all the credit.

Contents

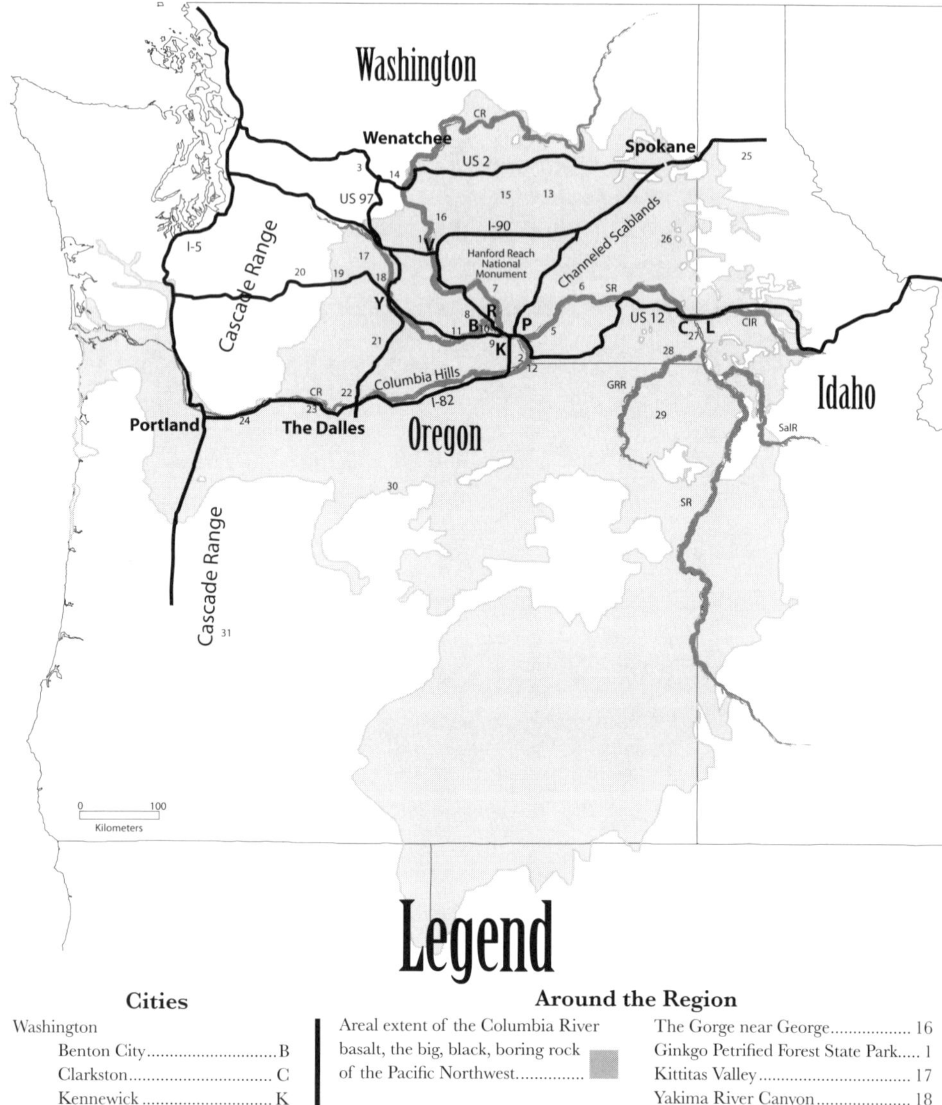

Legend

Cities

Washington

Benton City............................B
Clarkston...............................C
Kennewick ....K
Pasco ....P
Richland.................................R
Vantage V
Yakima ....Y

Idaho

Lewiston.................................L

Rivers

Columbia River............................ CR
Snake River.................................... SR
Yakima River.................................YR
Salmon River .... SalR
Clearwater River..........................ClR

Around the Region

Areal extent of the Columbia River basalt, the big, black, boring rock of the Pacific Northwest..............

Wallula Gap 2
Leavenworth 3
Ice Harbor Dam 5
Palouse Falls 6
White Bluffs...................................... 7
Rattlesnake Mountain...................... 8
Badger Mountain 9
Red Mountain 10
Badger Coulee................................... 9
Brown's Rock & the Badlands......... 11
Two Sisters 12
Odessa Ring Craters 13
Pinnacles State Park 14
Soap Lake.. 15
The Gorge near George.................. 16
Ginkgo Petrified Forest State Park..... 1
Kittitas Valley................................. 17
Yakima River Canyon.................... 18
Fife's Peak Volcano......................... 19
Clear Creek Falls............................ 20
Satus Pass 21
Maryhill Museum 22
Rowena Gap 23
Bridge of the Gods......................... 24
Lake Coeur d'Alene 25
Steptoe Butte.................................. 26
Swallows Rock................................ 27
Fields Spring State Park 28
La Grande Basin 29
Fossil 30
Grande Ronde Basin....................... 29
Mount Multnomah & Three Sisters .. 31

Preface

A drive across eastern Washington's Columbia Basin offers expansive views of an often windswept, barren landscape populated by an endless supply of basalt boulders and formations.

To most people, they're just big, black, boring rocks.

But Steve Reidel sees volcanic destruction that buried 70,000 square miles of the Northwest under the world's largest flows of molten rock. He sees lava flows that repeatedly changed the course of the Columbia and Snake rivers and wiped out salmon spawning grounds for millions of years.

He also sees massive windstorms that deposited 150 feet of dust in places and cataclysmic floods that devastated the landscape. Those forces also left behind a breathing 190-foot waterfall in the middle of the desert and deposited the foundation of a thriving wine industry.

That's just for starters. Move on down the highway with Steve and you will see earthquakes that ripped apart the earth and exposed massive fault lines. He even sees a possible meteorite shower that spattered a 60-square-mile area near Odessa, leaving behind some 100 craters up to 2,000 feet across. That's not to mention flip-flopping magnetic fields that at one time millions of years ago would have left your compass pointing southeast instead of north.

All because of big, black, boring rock.

In the fall of 1991, Steve responded to a classified advertisement in the *Tri-City Herald* newspaper seeking a columnist who would help residents of the Mid-Columbia appreciate this unique world around them. The idea was to describe the region's geology—and to do it in a way that wasn't boring—that would help readers easily understand the awesome powerful forces of nature that carved the face of eastern Washington.

As a geologist with Pacific Northwest National Laboratory, Steve became the Herald's "roadside geologist." He was directed to imagine himself sitting behind a steering wheel and wondering what forces formed that sprawling ridge or odd-looking pile of rocks. He's done exactly that.

With roughly 30 years of studying eastern Washington's ubiquitous basalt, Steve is well qualified. He sees the world as it was in the days when the course of rivers changed and mountain ranges virtually disappeared. He sees potential oil and gas locked under layers of basalt up to three miles thick, as well as a potential resource for "carbon sequestration," that could help combat the greenhouse effect. And he imagines the view that the Lewis and Clark explorers saw in 1805—and tells why that view also may offer a glimpse at the landscape of other planets.

For more than 13 years, Steve's essays have made geology approachable in verbal pictures punctuated by a dry sense of humor. A couple of examples, the first from a story during the Winter Olympics on the Northwest's record-holding basalt flows:

"A typical basalt flow erupts at nearly 2000 degrees. Summers may seem hot now, but nothing compared with temperatures in the Basin 17 million years ago."

And from a story on the far-ranging wanderings of the Snake River: "The salmon came back to the Snake, but Othello has not regained her river town status—yet!"

He's also wandered into the fanciful, as in this sentence from a column on the geology around the renowned music venue, The Gorge outdoor amphitheater: "The river still cuts, the land still rises, and this play undoubtedly will outlive all the players at The Gorge."

So read, then hop in your car and explore. You'll have a renewed appreciation for those black rocks, and maybe even be able to tell if they're basalt "bombs" or "spatter." You'll find out why Soap Lake is so salty, how Steptoe Butte is the sole survivor of a once vast mountain range, and how a fairly small volcano that you can walk to was one of the world's biggest lava belchers.

Boring rocks indeed.

Rick Larson, Managing Editor
Tri-City Herald
July 2005

Making and Shaping the Columbia Basin

Columbia River Basalt Lava Flows

No matter where you travel in the Columbia Basin, lava flows of the Columbia River basalt are a prominent part of the landscape. The largest known lava flows on Earth are the basaltic lava flows found here in our own little corner of the Pacific Northwest. These lava flows are known as the Columbia River Basalt Group.

Over 18 million years ago, a great plume of hot, molten rock coming up from deep within the mantle of the Earth was slowly spreading beneath the solid outer layer of the Earth's crust under the Pacific Northwest. This plume or "hot spot," the more common name, was centered under southern Oregon near the common boundaries with Idaho and Nevada. The hot spot was so large that its "head" spread almost 300 miles in all directions, reaching as far north as Washington.

Then, abruptly, about 17 million years ago, great cracks opened in the ground and the hot magma began pouring out on the land surface as basaltic lavas. This marked the beginning of the eruption of the Columbia River basalt lavas, which lasted until about 5 million years ago. They stopped only when the North American Plate, the large slab of rock that forms the outer layer of the Earth around us, moved away from the hot spot. That hot spot now sits under Yellowstone Park.

Basaltic lava is the most common rock in the world. Basaltic lava makes up the ocean basins and islands like Hawaii. Hawaiian lava flows are what most of us think of when we think of active volcanoes. Kilauea volcano on the "big island" of Hawaii has been actively erupting lava for decades. The latest Kilauea eruption began in January 1983. Since 1986, flows have traveled seven to eight miles from the vents to the sea, paving about 42 square miles of land on the south flank of Kilauea and building many hectares of new land.

With Hawaiian lava eruptions as our modern "measuring stick," we can compare them to the Columbia River basalt lava flows from the past.

The Columbia River basalt comprises more than 300 lava flows. The volcanic province covers over 70,000 square miles of the Pacific Northwest. Compared to the state of Hawaii, which is composed of many islands that are all basalt volcanoes and sitting in an ocean basin that is all basalt, our volcanic province is smaller. However, when you start comparing lava flows to lava flows, Hawaii pales by comparison.

Typical Hawaiian lava flows cover approximately 20 to 30 square miles with a total volume of much less that 1 cubic mile. Some of the Columbia River basalt lava flows cover over 50,000 square miles and consist of over 2,400 cubic miles of lava. There are no other known lava flows on Earth of that size.

To put this in perspective, the lava flows that you see as you drive through Spokane are the same ones that you see at Vantage, Wallula Gap, and on down through the Columbia Gorge. In Hawaii, the volcanoes are made up of thousands of individual lava flows. Here, ours are just a few really big ones.

A Columbia River basalt lava flow erupted in only a month or two at most. Between two eruptions, 20,000 years could pass when no eruptions occurred. However, when you start thinking about it, one of our lavas flowed from Spokane to the Pacific Ocean in a little over a month. That works out to one-half mile per hour.

If a lava flow started to erupt near Spokane, it would take about 10 days to reach the Tri-Cities and another 15 days to reach Portland, Oregon. As the lava moved westward, it would also be spreading toward Wenatchee and Lewiston, Idaho. In other words, these lava flows could cover half the Columbia Basin in less than two weeks.

The Hawaiian lava flows have created a lot of havoc in Hawaii. People have stood by helplessly as the flows destroyed their homes. With the size of the Columbia River basalt lavas, they would destroy everything in the Columbia Basin.

There is one last thing that you might want to think about. A typical basalt flow erupts at nearly 2000 degrees Fahrenheit. Think of the heat given off by that lava flow while it was cooling. If you think summers are hot now, I wouldn't have wanted to be here for a summer 17 million years ago.

Inside a Basalt Flow

The great lava flows that geologists call Columbia River basalt are the landmark rock of the Columbia Basin. Lately, they have been overshadowed by all the publicity given to the Ice-Age floods. Granted, the history and the landforms of the Ice-Age floods are an exciting story, but the basalt lava flows are just as fascinating with just as spectacular a history. Unfortunately, one reason the Ice-Age floods get better press is because people think that basalt is just a black, boring rock, and that all basalt lava flows look alike. That just isn't so.

The black rocks that bore many of my colleagues are the largest lava flows known on Earth. There are close to 300 individual lava flows, and some flowed out over as much as 40,000 square miles of the Pacific Northwest. Many lava flows extend from Spokane to the Pacific Ocean.

When you stop and look at a basalt flow along the side of the road, it may just look like a black rock, but there is a fascinating history hidden in there. You just have to know where to look for it.

The first thing you'll see in the black rock is a lot of fractures. The fractures aren't oriented randomly: there are two types. Fractures found in the lower part of the lava flow are long, straight ones that extend from the bottom of the flow vertically upward. These fractures are usually spaced several feet apart. When you look closely, you'll notice these fractures form long, straight columns with five to six sides. This portion of the lava flow is called the colonnade.

In the upper part of the lava flow, the fractures are more closely spaced and form a swirling pattern. This type of pattern is called the entablature. Here, the fractures are typically separated by four to six inches and also form five- to six-sided columns. Sometimes the swirling pattern gives the impression of a "war bonnet," and at other times, the swirling patterns intersect, giving all sorts of optical illusions. What you see is limited only by your own imagination.

A volcanologist named Tomkeieff first used these names for the fracture patterns while studying the Giants Causeway in Northern Ireland in the early part of the 19th century. The fracture patterns reminded him of classical Greek temples like the Parthenon. The Greeks carved images of their gods and the lives of their gods in the entablature, which sat on the majestic columns of the colonnade.

Like the Greek temples, the history of the lava flow is "carved" in the entablatures and colonnades. All the fractures formed when the lava flow cooled. Liquid lava is nearly 2000 degrees Fahrenheit when it erupts, and as it cools, it solidifies and shrinks. The fractures are really shrinkage cracks.

As the lava cools, it gives off heat to the air above and the rock below. Rock is a great insulator; the fiberglass insulation in a house is really made of rock. As a result, the lava cools more quickly from the top because air can carry away the heat.

The longer cooling time at the bottom of the lava flow allows the rock to form shrinkage fractures in the most efficient manner—large five- and six-sided vertical columns. However, in the upper part of the lava, the rapid cooling results in more chaotic column formation: smaller swirling columns.

As the basalt lava cools, the top and bottom of the lava flow become solid first, and the solidification progresses inward toward the center. The fractures follow the lava as it hardens, but fracture formation only happens in spurts. Only about two to six inches of a fracture will form at a time, even though the lava is continuously solidifying. Little "chisel marks" on the side of the fracture record each growth spurt.

Finally, after all the liquid lava becomes solid, the entablature and the colonnade meet at the two solidification fronts. One of our lava flows is typically about 100 feet thick, so it probably takes between 50 and 100 years for the lava to solidify completely, but another 100 years or more to cool to the point it is the same temperature as the surrounding air.

To a volcanologist, a pile of gravel left behind by the Ice-Age floods just can't compare to the geologic excitement of a cooling lava flow.

Pillow Lavas

Early scientists who first saw the Columbia River basalt lava flows in the Columbia Basin, saw many features they just didn't understand compared to classic flows like those of the Hawaiian islands. For one thing, our lava flows are 30 times thicker than a typical Hawaiian lava flow. A Hawaiian flow is about three feet thick; our lava flows are typically 100 feet thick. Columbia River basalt eruptions were so thick that many geologists thought that differences in features like the jointing patterns they saw meant they were really different lava flows. Some thought the upper, more densely joined part of the flow, the entablature, was a separate eruption from the lower, less densely jointed part, the colonnade.

One feature scientists saw they could relate to their Hawaiian experiences was a feature we call "pillow basalt." The name comes from the distinctive shape of the lava. Pillow basalts are discontinuous, elongate mounds of lava formed by the repeated oozing and quenching of the hot basalt by water. In other words, they look like rock pillows.

In Hawaii, when basalt eruptions flow down the volcano, they typically flow out into the ocean. There are spectacular pictures of the red, glowing lava as it enters the sea, and the water instantly flashes to steam.

As the basalts encounter the water, a flexible glassy crust forms around the newly extruded lava. Next, pressure builds from the flowing lava until the crust breaks, and new basalt extrudes like toothpaste, forming another pillow. This sequence continues until the eruption stops.

The pillows typically have an outer rim of black, glassy basalt that formed as the basalt chilled against the water. Inside the pillow, however, the lava is more crystalline because the insulating properties of the basalt allow the lava to cool more slowly and form crystals.

Pillow lava isn't the only thing that forms when lava enters water. The basalt shatters as it flows into water because of the lava's extreme heat: about 2000 degrees Fahrenheit. It's something like putting a glass you've heated in the microwave into a cold bath of water. The thermal shock just shatters the material.

The space between the lava pillows is usually filled with this shattered material. The shattered material is actually hydrated basaltic glass that is called palagonite. The glowing liquid lava freezes instantly as it enters the water, but a reaction occurs where the basalt and water combine to form the palagonite.

Palagonite is yellow, and when you look at the space between the pillows you see the shattered material is the yellow palagonite and pitch-black chunks of basalt glass. Geologists call the palagonite and basalt glass "hyaloclastite."

As lava flows into water and forms pillows, the pillow lavas and the hyaloclastics build out into the water, forming a delta in the same way a river will form a delta where it flows into a lake or ocean. You can recognize an ancient lava pillow delta by the inclined pillows. The delta forms in a progression as each new pillow layer is deposited on top of the previous inclined pillow layer. The delta will build out into the water, layer upon layer. Then, after the delta has been built, the lava flow continues to flow out across the delta forming a normal lava flow.

Pillow lavas and their deltas are very important to the Columbia River basalt. We see them in places like Vantage, Washington, and The Dalles, Oregon. The pillows tell geologists where rivers and lakes were, and more importantly, it's places like the pillow basalts at Vantage where the best petrified wood is found.

What Good is Columbia River Basalt?

When I started studying Columbia River basalt over 30 years ago, I thought there wouldn't be any commercial use for the basalt, other than storing groundwater and graveling roads. I remember one scientist at Washington State University (WSU) in Pullman who had investigated basalt as a source of insulation material. Rock is one of the best insulating materials available, and most of the insulation you can buy in a home-improvement store is made from rock. It usually is made from a silica-rich rock such as granite or sandstone that is melted and extruded into thin fibers woven to produce those mats you see at the store.

It turns out that basalt makes the best rock insulation material. However, the problem with basalt as an insulating material is that it melts at a very high temperature: about 2000 degrees Fahrenheit. The rock that goes into common insulation melts at about 1112 degrees Fahrenheit. This means that it costs a lot more money to make basalt into insulation, so the WSU professor's idea never became economical.

There is another possible use for basalt that my colleagues and I have been exploring at the Pacific Northwest National Laboratory (PNNL) in Richland for the past several years. This is to remove carbon dioxide, one of the greenhouse gases, from the stacks of coal or natural gas burning power plants and "sequester" it in the ground.

Carbon sequestration, as it is called, may be a useful means of dealing with excess carbon dioxide. It removes carbon dioxide from the stacks and pumps it underground, rather than letting it escape into the atmosphere. The U.S. Department of Energy's National Energy Technology Laboratory (NTEL) is funding many studies across the United States to see if this technology can help reduce greenhouse gases in the atmosphere.

Most of NTEL's efforts have been to examine deep saline aquifers that are usually found in rock like sandstone. The point is to get carbon dioxide out of sight. In the deep aquifers, the carbon dioxide is "sequestered" in water with the hope that it won't leak back to the surface. However, ask any hydrologist and they'll point out that there is no such thing as a non-leaking aquifer. A California national laboratory and the Texas geological survey did a field test in such a rock along the Gulf Coast.

The PNNL is looking at basalt as part of this carbon sequestration program because basalt has a property that no other rock has. If you're a mineral collector like the folks at the Lakeside Mineral and Gem Club, you already know why. It's because we find minerals in the basalt that show us that the basalt chemically reacts with carbon dioxide to form new minerals: calcite and siderite.

Calcite is calcium carbonate, and siderite is iron carbonate. Both minerals are a direct result of the reaction of the calcium or iron from the basalt with carbon dioxide from the atmosphere that is usually dissolved in groundwater. The product of this reaction is a new solid. That is the unique ability of basalt: it can turn a gas into a solid with a simple natural reaction. This means that the reaction can lock up carbon dioxide in a mineral, and the gas can't escape back into the environment.

One question is how long will the reaction take? If it takes too long, it may be able to escape anyway. If it takes too short of a time, then the place that carbon dioxide is injected could seal up right away.

My colleague has been carrying out experiments to see how long that reaction will take and under what conditions the reactions occur. As it turns out, he has found that it is not a fast reaction, so it won't seal the injection site, but it is fast enough so the gas can't escape back to the environment.

With the experimental work showing positive results, the next thing to do is a field demonstration. A field demonstration means drilling a borehole and injecting the carbon dioxide into one of the deep basalt aquifers. Most basalt aquifers deeper than 2,500 feet below the surface are so bad that they cannot be used for drinking water or irrigation so these are excellent test zones.

Even if a field test shows carbon dioxide can be safely locked up as a mineral in basalt, other problems remain to be solved. An economic technology to get the carbon dioxide off a power plant stack must be developed. It may turn out that, just like the WSU professor who had a better insulation using basalt, the financial costs of sequestering carbon dioxide in basalt (or anywhere) may turn out to be just too expensive.

Lewis and Clark Meet Columbia River Basalt

In 2005, we marked the bicentennial of Lewis and Clark's Corps of Discovery in the Pacific Northwest. Like all school children, I first learned of their expedition in American History, but now that I'm a geologist living and working in the heart of their discovery, their journey has taken on a whole new meaning. Reading their journals and looking at their maps is a fascinating adventure, especially when I fit their adventures into the geology of the Northwest.

Although the dams have submerged most of the rapids and islands that Lewis and Clark passed on the Snake and Columbia rivers, many of the features above the shorelines still can be found. Thanks to maps made before the dams, we can get a good idea of what the old rivers were like when Lewis and Clark traveled them.

One year my colleague, Duane Horton, and I had the opportunity to retrace some of Lewis and Clark's route down the Snake River with a group of teachers who were participating in the Pacific Northwest National Laboratory's Partnership for Arid Land Stewardship (PALS). We worked our way down the Snake River from Central Ferry to the Tri-Cities and compared the maps the Corps of Discovery made with maps the U.S. Geological Survey (USGS) made before the dams were built.

One thing we all learned from that experience was that Lewis and Clark were superb mapmakers. The comparison that the PALS teachers did between the two sets of maps showed that, considering the Corps of Discovery had instruments of the day and were traveling down the river in canoes, their maps were extremely accurate. The trend, shape, and width of the river on their maps matched our USGS maps so well that we could easily locate their landmarks.

However, there are places on their maps that are not as accurate as others. It turns out these places are usually where Lewis and Clark encountered rapids and, in some cases, very bad rapids. On October 16, 1805, the Corps of Discovery reached the confluence of the Snake and Columbia rivers. Up river, near Ice Harbor Dam, is one location where their map lost some of its accuracy. Their map indicates they encountered bad rapids and actually had to portage about three-fourths of a mile. Clark's journal said this area had "verry bad rapid or falls." Based on their map, this area had to be at Goose Island down river from Ice Harbor Dam.

For many years, we have been taking geology field trips to this area. Ice Harbor Dam is one of the few well-exposed volcanoes of Columbia River basalt. On the south side of the river is a craggy cliff at the end of the road west of the dam. This cliff is an 8.5-million-year-old volcano of basalt. Less than one-quarter mile down the river is another volcano that forms the next major cliff. Both these volcanoes were fed from below by basalt moving up through fissures in the ground. Geologists call fissures filled with once-molten rock a "dike."

At Ice Harbor Dam, the Snake River flows almost west. The basalt fissures that fed the volcanoes cross the river in a north-south direction. When we compared the geology to Lewis and Clark's map, the "verry bad rapid or falls" align with the location where the basalt dikes cross the river. It became very clear to us that these 8.5-million-year-old dikes created a major obstacle for the Corps of Discovery in 1805.

In our own investigation of the geology along Lewis and Clark's journey, the perils and hardships of that important part of Pacific Northwest history make us realize that even 200 years later, we're still on a journey of discovery.

Parallel Universes: The Basin and Mars

Believe it or not, many similarities exist between the Columbia Basin and Mars. The National Aeronautics and Space Administration (NASA) has long used the Columbia Basin as a model for designing various missions to Mars. Even today, planetary geologists study the Columbia Basin to try to understand what they see on satellite images taken of Mars.

Both Mars and eastern Washington are underlain by some of the largest lava flows known in the solar system. We call ours Columbia River basalt. Both have long, narrow ridges that extend nearly a hundred miles. We call these ridges names like Rattlesnake Mountain and the Horse Heaven Hills. The ridges of Mars are called "wrinkle ridges," but they're the same size and shape as the ridges of eastern Washington.

Our very own "channeled scablands" that were formed by the cataclysmic release of ice-dammed water during the Ice Age also have counterparts on Mars. Mariner satellite images show large valleys or channels that appear to be carved by great floods of water. The July 4, 1999, Mars Pathfinder set down in one of these valleys and confirmed many inferences made from earlier Mars Mariner missions.

Although NASA has had some failures, they have had some important successes as well. One mission, the Mars Global Surveyor, for example, found even more striking similarities between Mars and the Columbia Basin.

Jim Head and his colleagues from Brown University found on Mars what they interpret as a shoreline, which is consistent with an ancient ocean or lake. However, the data are not conclusive, and the matter of oceans on Mars is still a debate. In spite of this debate, there is little doubt in the scientific community that Mars had a lot of water in its past.

The water that carved the "channeled scablands" of Mars is thought to have existed about 3.5 billion years ago! Since then, the water disappeared, and now Mars is considered a frozen desert. It's too cold for water to exist on the surface and too cold for it to rain. In addition, the atmosphere of Mars is too thin for any significant snow to fall.

The question baffling geologists is: Where did all the water go?

Most scientists thought the water was just lost to space as it evaporated into the ancient Martian atmosphere billions of years ago. The Mars Global Surveyor made the case of the missing water even more interesting. At more than 10 locations on Mars, the satellite images show places where scientists believe water is seeping out of the walls of valleys and craters and forming small gullies. Some scientists think this may all be recent, as in the last 10 years. These water seeps suggest some of the water from past ages on Mars is underground.

If there is underground water on Mars, then we have another similarity with the Columbia Basin. We're a desert, and we too have water underground. Although we don't have a lot of it anymore, our underground water is the aquifer system that we rely on heavily for domestic use and irrigation.

One question I'm sure many of you may be asking is: So what? The answer lies in the question scientists have been asking for years: Is there life anywhere else in our solar system? One of the key ingredients necessary for life is water. If Mars has any chance of having life, there must be water there now.

There certainly is no life aboveground on Mars, but what about belowground? After all, on Earth we have all kinds of bugs and worms in our soil, but Mars doesn't have a soil zone like we do. Their soil is more like the permanently frozen ground of Alaska (permafrost) where bugs and worms can't live, especially if it's frozen year-round as on Mars.

Scientists from the Pacific Northwest National Laboratory gave the planetary science community an important insight to help answer this question. The answer, once again, came from the Columbia Basin.

In 1995, Jim Fredrickson and a colleague found bacteria living in the groundwater between the basalt lava flows over 2,000 feet beneath the U.S. Department of Energy's Hanford Site. They found that the bacteria survive not by using sunlight to make food, but by using inorganic carbon and hydrogen in the groundwater. The fascinating thing about their discovery is that the bacteria they found could probably survive on Mars.

Geologically Close to Mars

When Spirit and Opportunity landed on Mars and began exploring the two sides of the planet, Mars grabbed everyone's attention. Those two Rover mission probes were especially intriguing to those of us geologists who have been studying the Pacific Northwest for years.

NASA press releases said the search for water was the main goal of the two probes, but there were other parts of the mission that, although less well publicized, attracted as much attention among local geologists as the search for water.

Spirit used its RAT (rock abrasion tool) to cut a hole in the rock called adirondack, which looks like a chunk of basalt similar to our own. The close-up images appearing from the probes showed cavities, or gas bubbles, and large crystals set in the finer grained basalt. Geologists call these gas bubbles "vesicles" and the large crystals "phenocrysts." The phenocrysts are typically the mineral plagioclase feldspar.

Phenocrysts can tell geologists a lot about the history of the rock. Basalt first starts to form by the melting of a parent rock deep in the interior of a planet. Then the basalt magma must get to the surface, but often before it does, crystals may start to grow while it is in an underground storage chamber. Often, these first formed crystals and magma separate before reaching the surface. However, if they don't, the crystals show up as phenocrysts. So, when you see phenocrysts in the Martian basalt, you see evidence for geologic processes occurring deep in Mars.

In the Columbia Basin, you can see phenocrysts in fine-grained basalt if you drive out Horn Road along the Yakima River or along U.S. Highway 795 through Wallula Gap. In outcrops along the side of road, you can see little white plagioclase phenocrysts in the lava flows. You also can see similar RAT holes drilled by geologists who have studied the Columbia River basalt flows.

Geologists anxiously await the results of Spirit's chemical analysis of the adirondack rock. Chemical analyses can be used to tell how the rock evolved. If the phenocrysts separate from the magma deep in Mars, the chemical composition is altered. As more and more things happen to the magma, the composition progressively changes. One of the main changes is the amount of silicon dioxide.

Our own Columbia River basalts can have up to 56 percent silicon dioxide, which is much more than a typical basalt, such as Hawaiian basalt. The higher silicon dioxide content tells geologists that our basalts have had a long and complex history. Wait and see what adirondack has; if the silicon dioxide content is low, around 50 percent or less, you will know the Mars basalts haven't changed much from the original magma composition.

NASA named the local hills around the Spirit landing site the Columbia Hills. This was to honor the crew of the space shuttle Columbia who were lost in a tragic accident February 1, 2003. To those of us in the Pacific Northwest, this name has additional significance and makes us feel even closer geologically to Mars.

We have our own Columbia Hills. They form the north shore of the Columbia River from Patterson to nearly Hood River, Oregon. Our Columbia Hills and the Columbia Hills on Mars look very similar, especially if you compare NASA's picture to a view from south of Prosser toward Patterson along State Route 221. The only difference is the vegetation.

Geologically, it appears as if our Columbia Hills and the Columbia Hills on Mars are closely related. Our Columbia Hills are a long ridge of basalt that has folded and faulted. Ridges that look exactly the same occur on Mars. From the photos shown by NASA, it appears that the Columbia Hills on Mars and the Columbia Hills in the Pacific Northwest are twins.

As geologists have always said, remove the plants and animals from the Columbia Basin and you would think you were on Mars. Now let's see what Opportunity tells us about that part of Mars. If the layered rock there turns out to be the same as our layered sediments at the White Bluffs, I'll be looking for the Martian sagebrush and jackrabbits.

Ice-Age Floods get the Spotlight

The Columbia Basin is world famous for the great Ice-Age cataclysmic floods and the landforms they left behind. During the Ice Age, over 500 cubic miles of water from an ancient lake in western Montana—blocked by a temporary ice dam in the Idaho Panhandle—poured across eastern Washington.

The floodwaters scoured out the Channeled Scablands, and submerged the Tri-Cities under a lake up to 1,000 feet deep. After the floods drained through Wallula Gap, they scoured the Columbia Gorge and flooded the Willamette Valley before emptying into the Pacific Ocean where they cut huge submarine-like canyons many miles out to sea.

This cataclysmic flooding happened not once but many times. The last time being 10,000 years ago, which left us the Hanford Reach and many of our landforms. For years, the amazing story of the Ice-Age floods was known only to us in the Pacific Northwest and the geologic community. Much of the general public here and elsewhere had never heard of the floods.

If the National Park Service has their way, this may be about to change. For the last couple of years, the Park Service and a group of volunteers have worked together to inventory hundreds of prominent features along the floods' path in eastern Washington.

They developed a plan with several possibilities designed to increase public awareness about these incredible floods. One possibility would be to create an Ice-Age Floods National Geologic Region.

This park without boundaries would be the first of its kind in the nation and would use simple kiosks and signage on existing public lands to tell the story of the floods. All possible plans fully respect the rights of private-property owners.

The potential scientific and socioeconomic benefits to this area of such a national designation are as enormous as the floods. The Columbia Basin would get the national spotlight.

Tourists can taste Mid-Columbia wines and produce in the heart of this region because the products of the Ice-Age floods are everywhere around us, not just on an isolated piece of land.

My colleague, Bruce Bjornstad, is one of the recognized experts on the Ice-Age floods and has worked with our local group and others from the region. To those of us who make geology our life, this probably is one of the most important events to happen because it gives geologists an opportunity to show off the country we live in as one of the most exciting areas in the country or, for that matter, the world.

Terroir of Washington

Usually by Memorial Day weekend, the Columbia Basin is in full bloom. It's a great time to go wine tasting and contemplate local geology. It's probably not surprising to anyone that geology is important to Washington's wine industry.

A geologist friend of mine said it's all in the term terroir. Terroir, a term used mainly in France but becoming more popular in the United States, describes the complex relationship between geology, soil, climate, and other physical factors that influence the character and quality of wine. Terroir, viticulture practice, and the winemaker's skill combine to produce the great wines we see in our region.

Most of Washington's wine grapes are grown east of the Cascade Mountains in the Columbia Basin. Washington's wine growing region has several American viticulture areas or appellations. Red Mountain, the Yakima Valley, Columbia Valley, and Walla Walla appellations are the ones east of the Cascades.

Appellations are part of a system implemented in 1983 to identify U.S. wines in a fashion similar to the French system, *appelliation d'origine controlee.* Our appellations, however, fall under the jurisdiction of the Alcohol, Tobacco Tax and Trade Bureau and are defined mainly by geography. Unlike France, where the classification depends more on the type of grape in a geographic area, maximum production, alcohol content and such, an American appellation is defined by characteristics such as topography, soil type, geology, climate, and to some extent, historical precedent.

Many wine experts would argue that climate is one of the single most important factors in the quality of wines. That's probably true, but a great wine is the product of all the factors that contribute to the grapes and wines, plus the skills of the winemaker. For this reason, the word terroir is becoming a popular way of looking at wines.

The one common factor that ties wine to geology of the Columbia Basin is the Cascade Range. Our dry climate is a direct result of the rain-shadow effect produced by the high elevations of the Cascades. If climate truly is the most important factor, then you can thank the geology of the Pacific Northwest for providing it.

Even on a smaller scale, the local microclimates that set one appellation apart from another, one vineyard from another, or even one part of a vineyard from another are influenced greatly by local geology. The Red Mountain appellation has a southwest exposure that ensures sunlight even during the shorter days of winter when the sun is low in the sky. In contrast, much of the Walla Walla appellation along the Washington-Oregon border is just north of the Blue Mountains-Wallula ridge where it receives less sunlight, especially in the winter when the vines need sun to ward off the winter cold.

With not many exceptions, most vineyards in the Columbia Basin grow on Missoula flood deposits or sediment derived from them. The Missoula floods occurred during the Ice Age when over 500 cubic miles of water impounded behind ice dams was cataclysmically released across eastern Washington. These floods occurred many times during the Ice Age and are responsible for carving the channeled scablands north of us and leaving a blanket of sediment through much of the lower Columbia Basin. Each Missoula flood formed a temporary lake behind Wallula Gap as water escaped down the Columbia River. Sediment settled from these lakes and formed much of the soil cover below 1,200 feet.

After each flood, our Mid-Columbia winds blew the flood sediment around, leaving thick deposits of wind-derived sediment called loess on top of earlier rock and sediment. Periodic eruptions of Cascade volcanoes mixed volcanic ash with the loess, making a great soil for the grapes. The Missoula flood sediment from the lakes and the loess consists of silts and sand. As a result, water easily drains though the soils and provides excellent conditions for grapes to grow.

However, as anyone who has driven around the area knows, the geologic story is not as simple as what I've painted above. In some areas, the loess is thicker than others, and at other locations, only basalt bedrock is found. As ridges like Red Mountain grew, they shed basalt and produced coarse gravel that is interbedded with the loess, which can influence drainage. At other places, deposits of white caliche (calcium carbonate) resulting from our desert climate occur. Caliche can prevent water from draining and inhibit the vines from taking in nitrogen and potassium.

Complex is a word I often hear used to describe the character of a wine. It also is a good word to describe the geology of our wine-growing region as well as all the microclimates, soil variations, and other factors I can't even imagine. With all these complexities that are difficult to control, I never cease to be amazed by the skill of the winemakers to blend all the variables to make the great wines of our area. As my friend would say, on your next wine tasting trip, enjoy the terroir of Washington.

The Geology of Award Winning Wines

To illustrate the importance of geology in Washington's wine industry, we'll use a winner of a wine competition as an example: Gordon Brothers' 1984 Merlot. The Gordon Brothers' vineyards sit on a bluff overlooking the Snake River just north of Pasco at Levy Park.

The bluffs above the Snake River are made from our old friend Columbia River basalt. You can see two lava flows at this site. They are 10.5 and 12 million years old and came from volcanoes near the Washington-Oregon border. Overlying the basalt and forming the main soil in which the grapevines grow is a light-colored sandy soil with well-developed layers or what geologists call bedding planes.

This sandy soil overlying the basalt is called the "Touchet Beds" and was deposited by the Ice-Age floods that came down the Snake River. These are the same floodwaters that formed the waterfalls at Palouse Falls Park and many other features of the Columbia Basin.

When the floodwaters breached the divide between Washtucna Coulee and the Snake River at Palouse Falls, the waters turned south and followed the Snake River. As the floodwaters entered the Tri-Cities area, the only way out was through the Horse Heaven Hills at Wallula Gap. This water gap is very small compared to the volume of water that had to pass through it, so the water temporary ponded behind it, forming Lake Lewis.

Lake Lewis covered most of the Tri-Cities, but its water also spread up the Yakima River almost as far as the city of Yakima. As the water ponded, fine silts and sands settled out of it forming layers or "beds."

The northern shore of Lake Lewis was just north of the Gordon Brothers' vineyard and, in fact, most grapevines in the Yakima Valley grow in these same sediments. The vineyards at higher elevations are usually in the windblown deposits that started out as Touchet Beds in the valley bottoms.

The Touchet Beds, in addition to the climate and dedicated vintners, are one of the main reasons why vineyards exist here. The wine grapes, or vinifera, like sandy-silty soil that drains well and doesn't have a hardpan or calchie layer near the surface. Grapevines have deep roots—from 6 to 10 feet—that need to spread to gain nutrients, and they don't like to get their feet wet.

Generally, grapevines don't like alkali soils or thin soil covers over the bedrock. These are the caliche beds that produce the white coating on rocks or the hardpans and various "salts" that are left behind in the soil when the water-rich alkali metals sulfate and nitrate evaporate. Grapevines grow best in thick, rich soils without these salts. But some of the best wines are not always produced from grapes grown in these perfect soils. Often, vintners find the poorer alkali soils produce grapes with a flavor that produce wines of outstanding merit. Many of the cabernet wines of the Columbia Basin seem to be best when grown on alkali soils.

As you drive down the road to Levy Park that borders Gordon Brothers' vineyards, you can see a small pit along the road that exposes the Touchet Beds. Here you can see the sands and silts that help make the great wines of Washington.

Time in a Bottle

New Year's Eve is a good time to talk about geologic "time." Anyone who knows geologists knows we have a distorted sense of time. It's not that we're always late to dinner or we don't carry a watch, it's really a case of perspective. One clue as to why we think differently about time can be found in the rocks of the Columbia Basin. After all, the record of geologic time is the rocks.

Within an hour's drive of the Tri-Cities, you can put your hand on lavas that erupted 16 million years ago, plus or minus about a half million years. Lying right on top of those lavas are Columbia River sediments laid down between 3 and 8 million years ago. And, if that isn't bad enough, right on top are flood deposits that are only a few hundred thousand years old, give or take a few millennia.

The "amount" of time geologists commonly work with is beyond all normal comprehension. To most geologists, a 16-million-year-old lava erupted only yesterday, and a million years is just a drop in the bucket. Part of our affliction comes from something we picked up from the Earth. On a recent birthday, Earth was about 4.5 billion years old. However, its birth took over 500 million years, so that's kind of a relative birthday. And besides, with the age of the universe estimated at 14 billion years, our planet is just a new kid on the block.

As geologists, we deal with rocks, and every rock formed sometime in the Earth's history. Early geologists spent lots of time just figuring out which rock came first. This we call relative time. Relative time is like being on vacation for the summer and having someone collect your mail. When you get home, you know that a piece of mail on the bottom of the pile came first, but you don't know the exact day that it came. The exact age of a rock or the day a piece of mail came is absolute time. The early geologists really had no way of estimating the absolute age of a rock, much less the age of the Earth.

Because of the importance of the relative ages of rocks, the first geologic age dates were actually names of periods of time like Ordovician or Miocene. These names worked very well because they represented an unknown range of time on a relative time scale. To us old geologists, the names were the ages. After all, what's in a name? A rose by any other name would smell just as sweet...and be just as old.

With time, technology gave us the tools to determine the age of rocks. We soon learned that the volcanic ash in the road cut along Chemical Drive in Kennewick is about

13 thousand years old or part of the Pleistocene age, and that the Mount Spokane ski area sits on rock that is 90 million years old, which is part of the Cretaceous age (the age of dinosaurs).

The accuracy of geologic age dates varies, however. The 13,000-year age of the ash along Chemical Drive is probably within 1,000 years of its correct age, which is still within the Pleistocene. However, the age of the rock at Mount Spokane is probably only within 10 million years of its actual age. That still puts it in the Cretaceous.

It turns out that, the older the rock, the larger the uncertainty in its age. Conversely, the younger the rock, the more certain we are of the age. Laboratory equipment is used to determine a rock's age. It's simply the precision of the equipment that determines the uncertainty. As technology progresses, the age dates become more certain.

However, in a 90-million-year-old rock, does 10 million years of uncertainty in its age really mean anything to us today? To most geologists, the answer is no. Geologic processes seldom punch time clocks. We have precise dates like the time and day Mount St. Helens blew up because our lives revolve around hours and days. However, most geologic processes, like the deposition of the ocean sediments that make up Glacier National Park, occurred over many hundreds of millions of years. Rivers have been eroding mountains and flowing to the sea since there was water on Earth. Volcanoes have been erupting since the first millennium, whenever that was. Days, weeks, years, or even millennia mean nothing for most geologic events. Mother Nature is not this parochial. She only thinks of the big picture.

Because geologists deal with Mother Nature day to day, their perspective of time takes on the same dimensions. The mother teaches her children well. The day-to-day processes become viewed in the larger scheme of things, and an "event" can take place over millions of years rather than in hours, days, or years.

Now that you understand a geologist's perspective of time, you're probably wondering what we do on New Year's Eve? The answer should be obvious. Just sit back with Mother Nature and enjoy a good bottle of Washington wine.

The Horse: A Great Geologic Mystery

If you were to ask any school kid to make a list of things that symbolize the American West, the horse undoubtedly would be high on that list. In school, we are taught that the horse really isn't native to the Americas but was introduced by the Spanish during their age of conquest. All our wild horses are therefore descendants of these Spanish horses.

Actually, this is only half true. Paleontology, the field of geology that studies the fossil record, has shown that the horse can be traced back over 57 million years to the geologic period called the Eocene. The geologic evidence also shows that the horse really originated in the Americas and not Europe or Asia.

The oldest fossil horse is called Eohippus (Dawn Horse). It was small, about the size of a dog, and was graceful with a slender face and long tail. Eohippus ran on four toes rather than one toe like modern horses. Eohippus browsed, eating mainly leaves and shrubbery and not grasses like its modern descendants.

By about 35 million years ago, the horse had evolved to about the size of a sheep. It ran on only three toes rather than four like Eohippus. About 20 million years ago, when the great basalt lava flows that we now live on were erupted, the horse had evolved to be larger: it was now the size of a small pony. Its feet had three toes, but the middle toe was much larger, setting the stage for our modern horse, which runs on one toe. Also about this time, the horse made its first appearance in Europe and Asia. By the end of the Pliocene age, about 2 million years ago, the horse had evolved into our modern horse, Equus.

Horse fossils have been found in the Columbia Basin, proving that they once flourished here long before man. Although we don't know of any horse fossils in the sediments between the great basalt lava flows, fossil horses have been found in the sediment beds that form the White Bluffs on the east side of the Columbia River north of the Tri-Cities. These fossil horses range between 3.5 and 5 million years old. The fossils have been found in gravel beds near river level and just above the white ash bed about a third of the way down the bluffs. Geologists also think that the horse was roaming the Columbia Basin when the great basalt lava flows were erupted 6 to 15 million years ago and when the great Ice-Age floods occurred.

With this long history of the horse in North America, what happened to them? Why weren't horses roaming the plains when the Europeans conquered the Americas? Why are our only wild horses from the "old world" and not native?

We know that as little as 13 thousand years ago, long after the last ice sheet from the Ice Age had disappeared, wild horses roamed the American plains in great herds. Then, for some unknown reason, the horse became extinct in North America. It only survived in Europe and Asia.

Nobody really knows why the horse disappeared from North America. Some have suggested that a great epidemic like hoof-and-mouth disease or sleeping sickness swept through the herds. Others have suggested that, since their disappearance coincides with the appearance of early man in North America, man may have caused their extinction. The horse truly remains one of the great geologic mysteries of America.

The White Bluffs Fossil Record

We owe our earliest record of the Columbia River system to the massive outpouring of Columbia River basalt lava that started 17 million years ago during the Miocene Epoch. These lavas filled old drainage channels and forced the Columbia River and its tributaries to make new channels time after time. The lavas preserved most of the old channels, giving geologists one of the most complete pictures of how a river system evolved.

But, unfortunately, these same lava flows also produced conditions that left us with a scarce record of the life and climate of the times. It wasn't until after the eruptions ended in our area about 8 million years ago that geologic conditions changed, and abundant fossils were preserved. The post-8-million-year fossil record is best preserved in the White Bluffs and has proven to be a treasure trove of the life and climate from 8 to about 3 million years ago.

What we know about the fossil record from 17 to 8 million years ago comes from elsewhere in western North America and the world. The Miocene was the golden age of horses, and the fossil record in North America shows that ancestors of the modern horse, camel, rhinoceros, and elephant were probably roaming the banks of the Columbia River and countryside when the basalt was erupting. In addition, many extinct animals like the saber-toothed tiger were probably here as well.

One would think that with such a vast Columbia River system we would have had abundant fish fossils. Unfortunately, because of the devastating effect of massive basalt eruptions on the rivers, the fossil fish record in the Columbia River system from 17 to 8 million years ago is nearly absent.

However, fossil localities preserved in the Columbia Basin give us glimpses of the life and climate during the basalt eruptions. The Miocene forest at Vantage is one such location and is world famous for its petrified trees. No animals are preserved there. The petrified forest is located on the old Vantage Highway between Vantage and Kittitas, and it's well worth a half-day outing to see.

The gem of all fossil locales was found along the margin of the Columbia Basin by a chance discovery in 1972. At Clarkia, Idaho, a lake was uncovered that had formed when a river was dammed by one of the Columbia River basalt lavas 14 million years ago. Clarkia Lake, as it is called, was in the foothills of the Bitterroot Mountains where the mountains meet the Columbia Basin. The lake was about 25 miles long and probably about 20 to 30 feet deep.

Spectacular fossils are found in the lake sediments that show the area had a diverse and abundant population of plants and animals. Fish, plants, and even insects are preserved from 14 million years ago. The trees and plants from Clarkia Lake indicate that the vegetation on the hills surrounding the lake was similar to that of the forests of the hill-and-valley topography of the southern Appalachian Mountains. Mixed with those plants are plants from a climate similar to present-day East Asia. Although no ginkgo trees were identified, Metasequoia, an early ancestor of the redwood tree, was abundant.

In the higher elevations surrounding the lake, conifers similar to the modern western U.S. forests prevailed. Cypress trees similar to those found in swamps of the southeastern United States today dominated the lake. Insects that are typically rare in the fossil record abound in Clarkia Lake, making it a special fossil locality. Insects typical of the southeastern United States dominate the species found there. One insect name that I recognize is the caddisfly.

One of the most spectacular finds from the lake is the original coloring preserved in many leaves and insect fossils. The most impressive fossil I saw was a stinkbug that had its original coloring of iridescent orange and green.

Clarkia Lake probably lasted only about a thousand years, but the fossils tell us much about the Miocene climate of the Columbia Basin. We now know that 14 million years ago the area along the Idaho-Washington border was a warm, temperate climate. It was much more humid than today and had mild, short winters. Lake sediments suggest that storms were more frequent during the year.

Overall, it was a much warmer time than we see today. With predictions for a much colder winter than in the past, Clarkia Lake sounds like a very pleasant place to spend a winter.

Blowing Dust: An Important Geologic Legacy

If you drove State Route 240 from the Tri-Cities to Vantage after the June 2000 range fire on the Hanford Site, you probably noticed the warning signs for blowing dust. For a while, it seemed like they were a permanent fixture, but the grasses came back in the fall, and much of the problem was gone. By spring, perennial grasses once again turned the brown hills to emerald green just in time for a new fire season.

Photographs taken from the air over the Hanford Site since the fires, and examined by my colleagues at Pacific Northwest National Laboratory, show that the sandy soils are being sculpted by the prevailing winds. The seemingly flat landscape along the highway is showing low, pronounced streaking from the southwest and northwest, our main wind directions.

What is happening from the wind is just part of the never-ending story of winds and dust in the Columbia Basin. For at least the past several million years, dust has been a prominent part of the geologic story of our region. If it weren't for blowing dust, we wouldn't have one of the richest wheat-growing regions in the country. The Palouse Hills are where millions of years of blowing dust has accumulated.

In spite of the nuisance blowing dust causes us, it is one of our own important geologic legacies. Much of the surface of the Hanford Site and Columbia Basin has been repeatedly sculpted by the winds. Our geologic map of the stretch of land along State Route 240 seems to have two main geologic units: stabilized sand dunes and active sand dunes.

Some dunes have been stabilized for many thousands of years. An occasional range fire will burn off the vegetation, as it did in 2000, and some dunes will turn active until the next growing season. For the most part, however, much of the once-active sand dunes are now stable.

There is one part of the area that will always remain active sand dunes: the land between Rattlesnake Mountain and Ringold Coulee.

One of the nicest ways to see the active dune field is by taking a scenic jet-boat trip up the Hanford Reach. The dune fields are forever giving that part of the Reach a beautiful sand beach. This is quite a pleasant change from the rocky shoreline along most of the Columbia River. The jet-boat pilots love to give you a close-up view.

It's remarkable that the dune field begins just as you get to Ringold Coulee and ends when you pass the coulee. It's even more remarkable when you find that the sand dunes are only on the Hanford side of the river and not on the Franklin County side, where the coulee is. It seems odd that the sand dunes should closely follow the boundaries of the old coulee on the opposite side of the river.

After talking with my colleagues who watch wind patterns, I learned that the local geology influences the wind patterns, which, in turn, controls the active dune field. The valley north of Rattlesnake Mountain and the low notch Ringold Coulee makes in the White Bluffs combine to keep the dune field active.

As winds roar over Rattlesnake Mountain and through that valley from the southwest, they cause the winds to sweep across the Hanford Site. The notch carved in the White Bluffs northeast of Rattlesnake Mountain provides a low-elevation wind gap for the winds to blow out of the basin. The wind gap is downwind and channels the winds. This allows faster winds to move close to the ground before blowing out through the wind gap. The increase in wind speed is just enough to keep the dunes active and prevent the vegetation from stabilizing them as it does elsewhere on Hanford.

Minerals of the Northwest

The Pacific Northwest has some of the most diverse geology of any place in North America. To prove this, just drive around the state. Major mountain ranges dominate the west side of Washington, Oregon, and most of Idaho, and between these ranges are smaller mountains and broad plains.

With this diverse geology comes an additional dividend: a wide variety of rocks and minerals. The geologic processes that produce this varied landscape also produce many different minerals. Minerals are a geologist's key for understanding the geologic processes like mountain building or erosion.

Minerals have very specific chemical compositions, but many different minerals can have the same composition. Every mineral also has a range of temperatures and pressures where it can form. Geologists can tell the temperatures and pressures that formed a mountain range like the Cascades just by studying the minerals in rocks. Good examples of this are graphite and diamond. Graphite is made of carbon and makes a great lubricant or pencil. A diamond is also made from carbon, but try to write with it. The difference is that it takes a higher temperature and greater pressure to make a diamond. Graphite can form in the Cascade Range, but a diamond can form only very deep in the earth.

Minerals also have fascination; they have always captivated our interest. Just walk through any jewelry store and look at all the beautiful gems. A gem is just a mineral with some special property that makes it valuable to us. Usually, it is something like clarity, which just means it is clear, has an even color, and no flaws. Often, like in the case of a diamond, it is because of supply and demand. If there are few on the market, they cost a lot.

Most rocks around us are composed of just a few (probably less than 50) minerals. These are what we call the common minerals because they are easy to find. But there are many thousands of minerals in the world, and most of these are not very common. A mineral does not have to be a gem to be spectacular. Minerals fall into six basic crystal symmetries but Mother Nature can manipulate these into many beautiful shapes and sizes. To me, natural crystals are far more beautiful than a gem stone. And, they are rarer than diamonds and much more valuable.

I could go on talking about minerals but you really must see them to appreciate them. Take a field trip, even just a trip to a gem and mineral show. In the Tri-Cities, the Lakeside Gem and Mineral Club hosts a rock and mineral show at the Benton County Fairgrounds every spring. There are many great mineral displays to see, and club members are happy to answer all your questions about minerals and mineral collecting. They give demonstrations on collecting and displaying minerals and rocks at the annual show. If you want, you can even buy minerals and crystals without having all the fun of collecting them. This rock and mineral show has always been a great event to see many excellent examples of common and uncommon minerals and the beautiful crystals they form. If you're looking for a great family activity, this is it.

Gold

One weekend I picked up the *Tri-City Herald* and saw articles about high school and college graduations, and then the record unemployment and dismal economy. It didn't seem to be a good time for students, or anyone, to be looking for a job. I thought that students trying to earn money needed to find innovative ways to find work. Then the answer came to me—gold!

The price of gold has been doing well for the past several years. It has had its ups and downs but it has been over $360 per ounce for some time. Maybe gold prospecting is the answer to summer unemployment.

Washington has gold, not a lot, but some. There are two types of gold deposits—lode and placer. Lode gold is found in bedrock such as in the mountains. Placer gold is found in stream and river deposits.

One of the last operating gold mines in Washington, the Cannon mine, was a lode deposit and practically within the city limits of Wenatchee. Gold was mined from deep in the Earth until December 1994. When the mine closed, it wasn't for the lack of gold. It was because the mine was slowly working its way under the city of Wenatchee, and the town was afraid the city would collapse into the mine workings. If you ski at Mission Ridge, you have driven by the mine entrance.

Placer gold is free gold transported by water along with sediment. It may have started off as a lode deposit, but erosion freed it, and the streams did the rest. Placer gold has been found in most rivers, including the Yakima and Columbia rivers.

In the early 1800s, the Chinese worked many of the sand and gravel bars along the Columbia near Wenatchee. A very fine-grained gold like flour was mined for many years along the west bank of the Columbia River south of Malaga.

During the Great Depression, a gold dredge worked the gravels down river from Vernita Bridge. In those days, the price of gold was controlled and you could only get $20 per ounce. After 1934, the price was raised to $35 per ounce.

Many years ago, before his death, Francis Dvorak of Benton City told me that he and other men were able to earn $3 per day working the gravels on that gold dredge. That doesn't sound like much, but when you have no job and no other way to feed your family, you do what you need to do.

Also, there is an old lode gold mine that operated in our area. It was worked in the 1930s at about the same time as the gold dredge on the Columbia River. You can still see the old mine from Interstate 82 three miles east of Prosser. A reddish exposure just east of the end of the county road marks the old mine. The mine consisted of a 50-foot-deep shaft and an open cut about 100 feet long.

The Prosser mine was in a mineralized zone that formed along the contact between a basalt lava flow and sediment. The sediment was deposited by the ancestral Columbia River system about 10 million years ago. The gold and silver in the mineralized contact zone had to have been brought in along with the sediment (placer gold). The mineralized zone probably developed as the hot lava heated water in the sediment, and the minerals were concentrated along the contact in the pyrite.

Assays of the ore from the mine showed that there was silver there as well. The gold was in the mineral pyrite (fool's gold), and assays showed the average gold value was from 26 cents to $1.60 per ounce of ore rock. The silver assay ran about the same. However, one gold assay in 1934 showed $13.00 per ounce of ore rock. The mine was not profitable because the ore zone is small, but in the Depression days of the 1930s, it probably helped the owner survive the times.

So, in these tough times, here is an opportunity for all students looking for a summer of fun and profit. First, get a gold pan and find a nice place along one of the rivers in the Columbia Basin. Also, be sure to contact the Washington Division of Geology and Earth Resources for state regulations. Then start panning your way to prosperity. By the end of the summer, you should make enough money to pay for the sunscreen and insect repellent you'll need.

Columbia Basin Natural Gas

The high price of gasoline at the pumps is a subtle reminder that there is only so much oil in the world and, unfortunately, all the easy oil has been found. Almost everyone has a projection for how long it will be before we use up all the petroleum reserves on Earth, but it seems that, with the worldwide population explosion and all the emerging economies, it may be a lot sooner than we hope.

With all the easy oil having been found, especially in the United States, petroleum companies are now forced to look at more difficult targets that were ignored in the past. These more speculative areas are called frontier provinces.

We live in one of the largest frontier provinces in the United States: the Columbia Basin. What little U.S. exploration there is occurs almost exclusively in the frontier provinces where exploration is expensive, and the chance for a return on investments is not good.

Columbia Basin oil and gas exploration began in the late 1960s when Chevron drilled a wildcat well on Rattlesnake Mountain. The borehole went through more than 10,600 feet of basalt before Chevron gave up. Their target was the same source that produced methane (natural gas) for the only commercial gas field in Washington State, the Rattlesnake gas field just north of Rattlesnake Mountain.

The Rattlesnake gas field produced a little gas from its chance discovery by sheep herders drilling a new water well in the 1920s. It operated until World War II. The gas was shallow and trapped in a small hill that you can see in front of the main ridge.

Chevron wasn't looking for another Rattlesnake gas field. They were looking for a much larger trap below the basalt. Unfortunately, they never got through the basalt because no one really expected the basalt to be so thick.

In the early 1980s, Shell began the first major exploration program in the Columbia Basin. They drilled six deep exploration wells in Washington, one of which was more than 17,000 feet deep.

Shell's target was the sedimentary rocks below the basalt. Their geologists reasoned that the source of the methane was coal-bearing rocks like at Roslyn, Washington, along the western margin of the Columbia Basin. These coal-bearing rocks tilt under the basalt and the Columbia Basin.

Although everything indicated there was a source rock for gas and oil, a source rock is only one thing necessary to form an oil or gas deposit. In addition, the source rock must be subjected to the right temperature and pressure.

If coal is heated under just the right temperature and pressure, a chemical reaction takes place that causes some of the more volatile components of the coal to be turned into methane. Geologists call this coalbed methane, and it is an important source of natural gas in the western United States. The coal-bearing rocks are more than 15,000 feet below the surface, so the chemical reaction should take place.

Once you have a good source of oil or gas, there also must be a rock porous enough to hold the oil or gas and form a reservoir and a rock to trap the oil and gas, which is called a cap rock.

Shell reasoned that the big ridges of basalt would make good cap rocks, but they couldn't find a good reservoir rock. They had some encouraging results like at the BN 1-9 well, which flared a lot of gas during drilling, but their exploration ended in the mid-1980s without a major success.

Shell quit exploring, but not because of lack of encouragement. The price of gas and oil dropped drastically in the mid-1980s. The main factor in exploration is economics; that is, the price of crude oil and natural gas. If the price is low, then the cost of exploration doesn't warrant the investment, because there is little or no profit.

With the price of crude oil climbing again, there has been some renewed interest in the Columbia Basin. If the price remains high at the pump, there probably is a good chance you will see some oil and gas exploration rigs drilling once again. You see, petroleum companies are full of optimists.

The Birth of the Columbia River

The roots of the Columbia River system go back almost 1 billion years when all the land masses on Earth were joined together into one "super continent." This super continent began to break up about 800 million years ago when the tectonic plates that make up the outer layer of the Earth began to separate.

One place the super continent separated was here in the Pacific Northwest. All the land west of Pasco and Othello broke away, and the Pacific Ocean was born. Those former lands are now parts of Siberia and Australia.

With the birth of the Pacific Ocean, rivers started flowing west to the ocean. We really don't know much about any of these rivers, but we do know that, since that time, there have always been rivers in the Pacific Northwest bringing sediment to the ocean. These were the ancestors of the Columbia River and its tributaries.

The next 600 million years were relatively quiet times in the Pacific Northwest. The rivers here probably looked more like the Mississippi River than the present Columbia River. A large river delta built westward into the Pacific and, for the most part, the Pacific Northwest coastal plane looked much like the Gulf Coast today.

All this changed about 180 million years ago. The quiet shoreline of the Pacific Northwest came to an abrupt end. The large tectonic plate that carried the Pacific Northwest and Pacific Ocean broke along the edge of the continent. The part of the plate that carried the Pacific Ocean began to collide with the North American Plate.

This brings me to the theory of Plate Tectonics. The theory describes what happens when two tectonic plates collide. As the plates collide, one plate overrides the other forcing it down in the Earth to be consumed by heat and pressure. This is the ultimate in recycling.

You can think of this process as a conveyer belt moving boxes between two platforms. When the conveyor belt (tectonic plate) reaches the opposite platform, it disappears under the platform (the other colliding plate). Here, the Pacific Plate disappears under the North American Plate. Islands and small continents are the boxes that are carried along for the ride. But when the boxes reach the other platform, someone must remove them. In the case of Plate Tectonics, the islands aren't removed and simply stack up on the other colliding plate.

This is the process that built Washington westward from the Tri-Cities over the past 180 million years. Each time a new island arrived where the two plates met, the North American continent grew a little more. The islands became part of North America, and the shoreline moved west.

The westward growth of Washington reeked havoc on the ancestors of the Columbia River. The rivers continued to flow west to the ocean, but they had to find their way through all this jumble of rock, much of it in the form of mountains. The rivers could no longer simply meander across the land; they had to do some serious work: erosion.

By 40 million years ago, Plate Tectonics threw a new obstacle into the rivers' path: the Cascade Range, which started to form at that time. All the former channels that the rivers had eroded were now rising with the mountains. In an effort to counter the uplift, all the rivers east of the Cascades united and formed one channel through the Cascades; this was the birth of the Columbia Gorge and the Columbia River as we know it.

By 20 million years ago, the Columbia River and its tributaries were well established. The river system drained much of what we now consider the Columbia River drainage basin. Plate Tectonics was not beaten yet. By 17 million years ago, one last obstacle was presented to the Columbia River: the eruption of the huge basaltic lava flows that cover most of the Columbia Basin.

The lava flows turned out to be just a minor inconvenience, especially after what the river had put up with for the previous 800 million years. Besides, the Columbia River had no place else to flow. Its drainage basin was surrounded by high mountains, so it settled in for a long stay in the Inland Empire.

The Once and Future Columbia River

As long as there is earth and rain in the Pacific Northwest, there will always be a Columbia River. Volcanoes and mountains have long tried to destroy it, but they only succeeded in disturbing it. In spite of all the indignities it has suffered, the Columbia River still continues to carry the mountains to the sea, hoping always to become one with the oceans.

The one thing geology has taught us is that nothing ever remains constant. Given enough time, even the continents move. The future of the Columbia River is also change. Although we really don't know what changes await the Columbia River, it's always fun to speculate on what is likely to change and what may persist.

During the last 180 million years, North America has grown progressively westward. This will continue to be the fate of the Pacific Northwest for perhaps another 20 million years. Eventually, the collision between the tectonic plates underlying the Pacific Ocean and North America that is producing the Cascade Range will end. The plate collision will be replaced by the ever northward progressing San Andreas Fault.

Accompanying the end of the plate collision will be the end the Cascade Range. But with the death of the Cascades will be the birth of a new and larger mountain range west of the present shoreline accompanying the arrival of the San Andreas Fault.

Just as the Columbia and the rivers of western Washington finish off the Cascade Range, the Columbia River will have this new mountain belt to face. But, as history has shown, this is all in a geologic day's work. No mountain range has ever defeated the Columbia River, and no mountain range ever will. They don't call it the mighty Columbia for nothing!

At the headwaters of the Columbia River lies the Canadian Rocky Mountains. Slowly and persistently, the Columbia has been wearing these mountains away. Eventually, the river will complete its task, leaving a broad featureless plain much like that of the Midwest.

So, with the Rocky Mountains wearing away and a new mountain belt along the coast to look forward to, what else is in store for the Columbia? A little closer to home, the Columbia might expect a few local irritations but nothing it can't handle. Since the great lava eruptions of the Columbia River basalt began 17 million years ago, the Columbia constantly has fought to keep its channel in one place. Eventually, the Columbia outlasted these eruptions, but it only got a short rest. Local ridges like Rattlesnake Mountain soon began to impose on the Columbia. The river was able to defeat many of these ridges like the Saddle Mountains and the Frenchman Hills by cutting water gaps through them, but other ridges, like Umtanum Ridge near Priest Rapids Dam, appear to have won out and changed the course of the river.

However, like the Rocky Mountains, the local ridges really are no match for the Columbia River. Slowly the Columbia River will wear them away until it flows across the ridges as if they weren't even there—and they won't be because the river will have removed them. The big bend of the Hanford Reach and the sweep down through Wallula Gap and along the Columbia Hills to the Columbia Gorge will be just history. It may take 50 or 100 million years to write this history but, nevertheless, it will be written.

The newest and most vulnerable part of the Columbia River, at least for the near term, is the Hanford Reach. The Reach was born at the end of the last Ice Age and is here because of the Ice Age. Most geologists believe that we are between ice ages, and the next one will begin in about 20,000 years. If this is true, then the Hanford Reach may have a short life for the fates that control rivers might well decide to bring the Columbia across the Hanford Site once again at the end of the next Ice Age.

You might think the Columbia River is immortal. And it is, as long as rain and snow continue to fall on the Pacific Northwest.

River Town of Benton City

If you were a resident of Benton City 8.5 million years ago, you would have been living along the banks of a large river. Now the first thing you might say is, "Wait a minute, Benton City is along the banks of a river, the Yakima River." There's only one difference, however, the Yakima River flows from the Cascades past Benton City, but this earlier river was flowing toward the Cascades.

The river flowing west past Benton City 8.5 million years ago was the Columbia River and not the Yakima River. From 8.5 to about 5 million years ago, the Columbia River flowed along the same course in the lower Yakima Valley that the Yakima now uses, but in the opposite direction.

In the last 15 million years, the Columbia River has moved slowly but steadily eastward from the edge of the Cascades to its present location. Its path through Benton City and the Yakima Valley is just a small part of that history of change.

For much of the early history of the Columbia River, it followed a course that roughly coincided with U.S. Highway 97 from Yakima to Goldendale along the east side of the Cascades. At that time, the edge of the Cascades was the lowest point in the Columbia Basin. However, that was to change as the ridges that surround us began to grow and gain elevation to the west. This process caused the Columbia River to be forced slowly eastward.

About 10 million years ago, the Columbia followed a path that took it from Sentinel Gap just south of Vantage to Vernita Bridge and the west side of the Hanford Site, then through the Sunnyside Gap Highway (now State Route 243), and on down to Goldendale and the Columbia Gorge.

Around 8.5 million years ago, two events conspired to force the Columbia River into the lower Yakima Valley. The first was the eruption of one of the large lava flows of Columbia River basalt that dominate our region. The lava erupted near Ice Harbor Dam and made it just as far west as the current course of the Yakima River, where it flows from Rattlesnake Mountain to Benton City.

The second event was the continuing growth of local ridges, Rattlesnake Mountain, in particular. These ridges were growing higher to the west of us, and rather than fight with the ridges, the Columbia River took the easy way and shifted its course eastward.

As the Columbia River shifted eastward from Sunnyside Gap, it encountered the edge of one of the Ice Harbor lava flows near Horn Rapids. The lava formed a barrier that prevented the river from moving any farther south. As a result, the river flowed along the edge of the lava to Benton City and the Yakima Valley.

The Columbia River followed this course for nearly 3 million years until uplift caused the upper Yakima Valley to reach higher elevations than Benton City at the end of the valley. With this change, the Columbia was forced out of the Yakima Valley and started to flow through Wallula Gap. This change also marked the time when the Yakima Valley began draining the Cascades and the Yakima River was born.

Evidence of the ancestral Columbia River in the Benton City area abounds. If you drive the Old Inland Empire Highway just west of town, you can see a prominent bench that sits above the cliff overlooking the river. Along this bench, you can find gravel composed of many exotic rocks from northern Washington and Idaho. These exotic rocks are important signs to geologists that the Columbia River was once here. This bench was carved into the lava flow that predated the 8.5-million-year-old Ice Harbor lava. This bench is even more prominent on the south side of the river where many new vineyards have sprung up.

Geologists have traced this bench and gravel deposits up the valley toward Prosser, but the old valley gets wider west of Chandler. Geologists even found these ancestral Columbia River gravels on top of the Horse Heaven Hills along McBee Grade out of Benton City. The McBee Grade gravels show geologists just how much the Horse Heaven Hills have grown in the last 8.5 million years and how much the Benton City area has been influenced by its geologic legacy.

Drought and Groundwater

We expect it to be dry in our area. The Columbia Basin is called a desert because we receive less than 6 inches of precipitation a year, so typically we can't count on rain or snowmelt for water. We rely on the Columbia River for drinking water and irrigation, but we also rely on groundwater.

If you don't live near a river, chances are your house uses groundwater. The Yakima Valley, Othello, and rural Benton and Franklin counties all depend on groundwater for drinking and other domestic uses. Many farmers who irrigate also count on groundwater for their livelihoods.

The Columbia Basin overlies several sources of groundwater that people have been using since the land was settled. The water is found mostly between the lava flows of Columbia River basalt, but close to the Tri-Cities, there is water in the sediment above the lava.

Groundwater always appears like a mystery to anyone trying to find it. It seems as if you can put in one well that is dry and then another only a few hundred yards away and get more water than you can possibly use. This isn't random luck; groundwater is not controlled by the fates.

Geology controls groundwater, not the "god of the underworld," as a colleague once remarked. The more you know about the geology of an area, the more you know about groundwater and where to find it or not find it.

How do you get groundwater? The first thing you need is large amounts of water that pass slowly into the ground. This is called recharge. Recharge typically occurs at higher elevations where snow collects and can slowly melt and seep into the ground. Much of our groundwater originated in the mountains that surround the Columbia Basin and from the great Missoula floods. However, some of our groundwater comes from irrigation practices.

Next you need a rock that will hold the water. As you might expect, rock doesn't hold much water; that is why we have rivers and streams. Groundwater occurs in the very small open spaces in rock called pore space and in fractures in the rock. There are no underground rivers. The more pore space or fractures in the rock, the more water you have a chance of getting out.

But just because the rock has many small pore spaces or fractures doesn't mean your water well will pump 400 gallons per minute; you may not get any water. The small pore spaces and fractures must be connected so water can move from one pore space or fracture to another. This is the permeability of a rock. Water has to be able to get from where it enters the ground to where you want to pump it out. With a high permeability and a lot of water, you might have a good water reservoir, or aquifer, as geologists call it.

Having sufficient water is a serious problem. Just ask anyone drilling a well or someone who has had to deepen their well because they were running out of water. Unfortunately, in the Columbia Basin, we are using the groundwater much faster than we are replacing it. This means we are going to run out someday. At the rate we are using it, that probably will be sooner than later.

Besides abundance, unfortunately there is another problem with groundwater: quality. Water quality is a problem many people don't think about. Just because your tap water is clear doesn't mean you have high-quality groundwater.

We rely on recharge to replace the water we use. Unfortunately, the water that recharges our aquifers moves from the surface down. That means any contaminant or chemical in its way can be carried along and give us poor-quality groundwater.

Abundant, high-quality groundwater is a worldwide problem. For that reason, water is said to be the oil of the 21st century. On hot days, water conservation and protection are an important investment in our future and the future of generations to come.

Cattails in the Desert

It's no surprise to see sagebrush, rabbitbrush, and other desert plants covering the countryside in the Columbia Basin. And we expect all the lush, green vegetation to be in narrow strips along rivers and in places like the Potholes Reservoir. But we're not used to seeing an oasis out in the middle of the desert away from any obvious source of water.

The Columbia Basin has its share of oases where cattails thrive. One good example is in Badger Coulee, south of Kennewick. Badger Coulee is typical of the large, dry valleys in eastern Washington. Although there is no river or even small stream flowing through the coulee now, Badger Coulee was once the main channel of the Yakima River.

The Yakima River cut this channel millions of years ago as it flowed southeast from Benton City to join the Columbia River at Kennewick. It was forced into its present course when the entrance to the coulee at Benton City was blocked by a gravel deposit left behind by the cataclysmic floods of the Ice Age. This is now the A&B Asphalt Co. gravel pit.

Badger Road winds through the heart of Badger Coulee and is a pleasant alternative to the freeway between Kennewick and Benton City. To see this oasis, drive south on Badger Road from the Clearwater/Badger Road interchange on Interstate 82.

As you leave the freeway interchange, almost all the plants are typical sagebrush country vegetation. If you look to the right (west), you'll see a few trees growing along the ephemeral stream that carries away rainwater and snow melt in the winter and spring. If you were to walk over to the trees in the summer and fall, you wouldn't see any water. The tree roots take up groundwater to survive.

About 1.7 miles from the freeway, you'll notice the valley becomes green. In fact, cattails grow along the right side (west) of the road. You'll find cattails and Russian olive trees growing for some distance ahead. For a long time, this area has supported plants that require a lot of water. When geologists see something unusual like this, it gets their curiosity going because there must be some interesting, underlying geologic reason.

In Badger Coulee, the clues to this puzzle are all around. First, look to the left (east) and you'll see an irrigation canal that runs the length of the coulee. For a long time, the irrigation district took all the blame for this wetland. A leaking irrigation ditch was blamed, but no one could ever find the leak, and the wetlands were there before the irrigation ditch!

Next, look around at the surrounding hills. One of the first clues is the change in topography along the rim of the coulee. As you come south from Kennewick, you'll see the relatively flat, uniform rims of the walls of the coulee. Where the cattails begin, the rims of both coulee walls to the south abruptly become several hundred feet higher. This is caused by the Badger Coulee fault. This fault formed as the Horse Heaven Hills rose to the south. The fault parallels the Horse Heaven Hills and cuts right across Badger Coulee where you first see the cattails.

If you live outside the city limits, chances are you drink groundwater. Our major sources of groundwater are the aquifers that occur between the lava flows that make up the bedrock of the Columbia Basin. If you want to use groundwater from one of the aquifers, you have to drill an expensive water well at least several hundred feet deep. In Badger Coulee, however, Mother Nature saved the residents the expense. She simply lifted up one of the best aquifers in our region as the Yakima River cut down through the lava flows.

Today, south of the fault, this aquifer flows directly out into the coulee, but you don't see the water because the aquifer is covered by gravels. The gravels are thin enough, however, by the Badger Coulee fault that the groundwater is right at the surface, giving us an oasis and cattails in the desert.

Landmarks

Rattlesnake Mountain

If someone were to ask you to name the most important landmarks in the Columbia Basin, I bet Rattlesnake Mountain would be near the top of your list. Rattlesnake Mountain is probably the most distinctive landmark in the Columbia Basin. Everyone knows it. No matter what direction you drive, Rattlesnake Mountain always seems to be within sight. This is especially true when you drive from the east: the mountain stands out on the horizon long before you're anywhere near the Tri-Cities.

Rattlesnake Mountain is the highest point on the Hanford Reach National Monument, rising slightly more than 3,600 feet above mean sea level. This makes it several hundred feet higher than Snoqualmie Pass in the Cascades. From the highest point on Rattlesnake, the ridge drops off in all directions, reaching its lowest elevation of about 450 feet where the Yakima River cuts a water gap through its southeastern part.

Rattlesnake Mountain is typical of the ridges that make up the western part of the Columbia Basin. Geologists call these ridges the Yakima Folds because the rock has been folded into these distinctive arches that geologists call "anticlines."

Rattlesnake Mountain is made up almost entirely of basaltic lava flows that erupted millions of years ago from volcanoes farther east in Washington, northeast Oregon, and western Idaho. These are the same black rocks that cover almost all the Columbia Basin.

One look at Rattlesnake Mountain and you'll notice it's not very symmetrical. Along its length, it looks like a series of "steps" coming up from the Yakima River. In addition, the north side forms a cliff and is much steeper than the south side. The asymmetry comes from the tilt of the lava flows when they were folded. The lava flows are tilted more steeply on the north side than on the south side. The south slope is formed from the surface of just one lava flow and shows how the Earth's forces bent this rock.

On the north side, however, many lava flows are exposed along distinct breaks called faults and separate blocks of unbroken rock. These faults are easily seen because they are more eroded than the unbroken rock. The distinct break in slope that forms the base of Rattlesnake Mountain is the Rattlesnake Mountain fault. It is the place the rock broke as it was folded.

The Rattlesnake Mountain fault can be traced from the Yakima River for about 14 miles to the northwest. A second, smaller fault can be seen as a notch above the

Rattlesnake Mountain fault. This fault can only be found along the highest part of the mountain; that is, on the highest of the "steps" along the ridge. The two faults become one along the southeast part of the mountain where the mountain "steps down" in elevation toward the Yakima River.

Rattlesnake Mountain and the rest of the Yakima Folds have received a lot of attention from geologists in recent years and are starting to receive even more attention now. It's not because of the great beauty of these mountains, although beauty is in the eye of the beholder, but because they may be clues to the origin of our solar system.

Planetary scientists have seen features similar to Rattlesnake Mountain on the images of other planets sent back by NASA's space probes. Long, linear ridges called wrinkle ridges—now known to occur on Mercury, Venus, Mars, and our Moon—look just like Rattlesnake Mountain and our Yakima Folds. The wrinkle ridges are found only where huge lava flows like the ones that formed the Columbia Basin erupted onto the surface of the planet. Planetary scientists think the processes that formed Rattlesnake Mountain may hold the clue to understanding how other planetary surfaces have evolved.

The Story of Brown's Rock and the Badlands

Like many other kids growing up back East, I fell in love with the West through old classic westerns on TV like Hop Along Cassidy, the Lone Ranger, and Gene Autry. Some of my most vivid memories of those programs were the scenes where the bad guys would ambush a stagecoach in some narrow canyon or mountain pass or shoot it out on a mountainside where there were plenty of rocks to hide behind. All us kids were convinced that was exactly how the West was.

I soon learned that shootouts weren't the real West. However, I was happy to find that the scenery shown on TV was real, and it was the West's landscape that attracted many geologists from the East to this region and has kept us here.

It was my wife, Mimi, who made me realize how some of the old westerns I grew up with really were tied to the land. Once, Mimi told me a story her father told her when she was a little girl on a farm in Benton City. It was the story of Brown's Rock, which is located along the Old Inland Empire Highway that runs from Benton City up the Yakima Valley. Maps from the late 1800s show the highway was a stagecoach route. It also was a major trail probably used by so many generations of Native Americans it seemed like it had been there forever.

About 5 miles west of Benton City, the Old Inland Empire Highway crosses what geologists call The Badlands. This is a fairly large and rugged ravine just after the road climbs off the floodplain. Just west of The Badlands, the road weaves through a lot of huge boulders that have fallen from the cliffs above. This is the Brown's Rock area.

As the story goes, highwaymen would hide behind one of those large boulders and rob stagecoaches. The story also included a shootout, but the exact details are not well known.

This story does remind me of the TV shootouts and robberies in the westerns I watched as a kid. Now, with 40 years of geology behind me, the geology of the places the bad guys chose to ambush stagecoaches, or where they ultimately met their end, is as interesting as the stories themselves.

The geology of the Brown's Rock area, for example, is every bit as fascinating as the landscapes in those old westerns. The story incorporates every facet of the geologic history of our region. It involves the great basaltic lava flows, the Columbia River, and the Ice-Age floods.

The geologic storyline begins with The Badlands, which geologically are part of the Horse Heaven Hills. They formed as the Earth's forces compressed the Columbia Basin from the north and south. The basalt lavas of The Badlands were folded into a small arch. This small arch continues to the northwest as a series of small hills all the way up to the Rattlesnake Hills.

The Old Inland Empire Highway cuts through the youngest lava at The Badlands—which is 10.5 million years old—and the sandstone that lies below it. This sandstone is old sediment deposited by the Columbia River long before it flowed in its present channel.

The Ice-Age floods were the final geologic event that created The Badlands. Even though the main floodwaters passed through the Tri-Cities, some water surged up the Yakima Valley, cutting away rock and depositing sand and gravel like what you find at Whan Road on the Old Inland Empire Highway.

As the floodwaters roared up the valley, they stripped away rock from The Badlands to carve out the ravine and ripped out the sandstone to create the ledge the road follows. As the floodwaters removed the sandstone, they undercut the basalt as well. The unstable basalt then broke off along joints and fell to the bench below. One of these fallen rocks is Brown's Rock.

The Two Sisters

The Two Sisters rocks along the Columbia River at Wallula Gap are one of the Columbia Basin's oldest landmarks, a product of 15 million years of geologic history. According to a Native American legend, the two sisters were wives of Old Coyote, a spiritual hero of many legends. Coyote became jealous of the sisters and turned them into stone. But, a chronicler of the Lewis and Clark Expedition, Albert Salisbury, called the rocks the Two Captains as a tribute to Lewis and Clark.

The geologic story of the Two Sisters began with the eruption of Columbia River basalt, the series of lava flows that covers most of eastern Washington. The Two Sisters are just one of a series of flows called the Sand Hollow flows, named after Sand Hollow on the east side of the Columbia River near Vantage.

The Two Sisters and their base are part of just this one lava flow, but the two parts have distinctly different appearances. The difference is caused by the density of cracks that formed in the lava as it cooled. The base of the Two Sisters is made of basalt with large vertical columns. Geologists call this the colonnade of the lava flow. The sisters themselves are made of basalt that has more closely spaced cracks. This is the entablature of the lava flow.

Cracks in the colonnade formed from the base of the flow upward. Cracks in the entablature formed from the top down. These two parts of the lava flow are a result of the rate at which the lava flow cooled. Slow cooling causes the colonnade and more rapid cooling causes the entablature. The entablature is usually in the upper part of the lava flow because it is exposed to the air. The base of the flow is better insulated so it cools more slowly.

After the basalt flows cooled, forces in the Earth folded the basalt. As the basalt folded, it also fractured, and even larger cracks formed in the rock. These larger cracks are vertical and penetrated the entire rock. They stand out now as the vertical indentations in the Two Sisters.

The great Ice-Age floods did the final sculpturing of the Two Sisters. The floodwaters that dug the coulees and left behind thick gravel deposits could only drain from the Columbia Basin through Wallula Gap. The water pushing through the gap had so much force, it widened the gap and carved the Two Sisters. The floodwaters removed the cracked and fractured rock at the top of the sisters, but the large columns that formed the base were strong and resisted erosion. Erosion along one of the large vertical fractures probably was responsible for the separation of the Two Sisters. The Two Sisters probably have changed very little since Lewis and Clark saw them.

The Two Sisters are part of the Walla Walla County Parks system. They're located on Washington State Route 730, 2 miles south of U.S. Highway 12 between Pasco and Umatilla.

Badger Mountain

The Badger Mountain nature preserve has become a reality. This is a compliment to the Friends of Badger Mountain, government officials, and the landowners who were able to work together to make sure that another Tri-Cities landmark remains for future generations to enjoy.

Why should Badger Mountain be preserved? For one reason, it provides a small island of nature in the heart of our rapidly developing area. Within a short distance from home, a hiker can see native plants and enjoy a scenic view of the Tri-Cities as well as several Cascade volcanoes.

And, of course, the Badger Mountain nature preserve provides a wealth of geology to enjoy as well. Although much of the rock is covered by the soil and plant community, there are many geologic features that you still can see.

Badger Mountain is one of a series of northwest-southeast ridges along the Rattlesnake Mountain-Wallula Gap trend. These ridges are part of what geologists call the Olympic-Wallowa lineament, an alignment of geologic features that extend from the north side of the Olympic Peninsula to the north side of the Wallowa Mountains in Oregon. If you looked at a satellite image of the Pacific Northwest, you couldn't miss this feature.

It will probably come as little surprise that lava flows of Columbia River basalt make up Badger Mountain. The ages of the lavas exposed are anywhere from 14.5 to 8.5 million years old. Layers of sediment are found between the lava flows, but they're harder to see, except in the spring just after the wildflowers have bloomed and most vegetation is beginning to turn brown. Then, you can easily find some of the sediment layers by locating the greener plant cover. The sediment layers hold water and, thus, the plants that grow on them stay green for just a little while longer.

You also can tell a lot about Badger Mountain from its shape. The rock has folded so that it looks like a small rock dome sitting above the surrounding flat rock. Contrary to what you would expect to see in flat-lying rocks, the youngest lavas make up the lower elevations and the older rocks the higher elevations. This is because of the steep tilt of the rocks. As the rocks buckled, and the ridge grew higher, erosion removed the younger rocks from the upper elevations so now the oldest rocks are exposed at the top. The 12-million-year-old Pomona lava flow makes up the peak of the ridge.

The north side of Badger Mountain has the most complex geology on the ridge. Like all ridges in the area, geologists have learned from mapping the rocks that the north-facing slopes are fault zones. On Badger Mountain, there are two faults on the north side that merge into one larger fault on the northwest and southeast ends of the ridge. The main fault is at the bottom of the ridge.

If you start hiking up from Shockley Road, you cross over the fault just where the steeper slope begins. As you cross the fault, the rock that makes up the slope is tilted steeply to the north; it stands nearly vertical in some places. If you follow the gully on the east side of the small spur above Shockley Road, you pass through three lava flows: the 8.5-million-year-old Ice Harbor lava, the 10.5-million-year-old Elephant Mountain lava, and the 12-million-year-old Pomona lava.

When you get to the top of the gully, at about 1,300 feet, you have reached the upper fault. It's easy to find because the rock broken by faulting and weathering has produced a bench on the slope that can be traced to the northwest. The fault is exposed along Interstate 182 where the fault and rocks are tilted to the southwest, and several of the lava flows have been thrust on top of themselves.

As you continue your ascent to the top, you pass into the oldest rock exposed anywhere along Badger Mountain. The rocks no longer are steeply tilted to the north. By the time you reach the crest of Badger Mountain, the rock tilts gently to the southwest.

So, for those of you who like to hike Badger Mountain, don't just stop and smell the wildflowers, look at the rocks along the way; there is a lot of beauty there as well.

Ice Harbor Dam Volcano

You already know that Columbia River basalt is a prominent part of the local landscape. Rarely, however, do you see a volcano for these lava flows. That's because the lava for these eruptions came from fissures: narrow fractures in the ground that were 90 miles long and 100 feet wide. Because there were more than 300 eruptions, most fissures and their volcanoes were buried by younger eruptions. Even if the lava flows had not been buried by younger lava, to be seen today, they had to withstand millions of years of erosion. Nowadays, you can see these fissures only in eastern Washington and Oregon and western Idaho, where rivers cut deep canyons to expose them.

In the Mid-Columbia region, several volcanoes and basalt-filled fissures (called dikes) survived the test of time. You can see them near Ice Harbor Dam. These dikes and volcanoes are from the youngest basalt eruption in the area: the Ice Harbor lava flows, which erupted about 8.5 million years ago.

A well-preserved volcano can be seen on the southeast side of the Snake River on the downstream end of Ice Harbor Dam where the flat gravel area narrows at the river (about a mile from the dam). Here, the craggy bluff is a remnant of a volcano for the Martindale flow, the second of the Ice Harbor flows to erupt. Erosion has removed part of the volcano, exposing the inside.

This volcano is relatively small, only about 650 feet wide and 150 feet high, and is probably only one of many volcanoes that grew along the length of the fissure. It is composed of tephra (from the Greek ash). Tephra refers to volcanic material thrown into the air during a volcanic eruption. Near the base of the Ice Harbor Dam volcano is tan-colored tephra made up of fine volcanic ash particles and larger blocks of basalt, called bombs. Some of the bombs are pieces from older buried lava flows that were ripped loose and brought to the surface by erupting lava. On top of the tan tephra is a dark basalt tephra called spatter: large pieces of lava that were thrown into the air during the eruption, fell to the ground, and cooled near the fissure. This process built up the sides of the volcano. Overlying the tephra is lava that flowed from the volcano. The tephra is from the early part of the eruption that had lots of gas (water vapor and carbon dioxide.) The lava is the main eruption but with little gas.

If you walk down the path for a half a mile to the next craggy bluff (a small quarry), you will see a dike cutting the Martindale lava flow. The dike stands out as a rubbly 10-foot-wide zone cutting up through a lava flow from the volcano you just saw. The lava flow has distinct vertical fractures. This dike is the fissure that supplied lava to the youngest Ice Harbor lava flow, the Goose Island flow.

The type of eruption that formed these volcanoes is called "Hawaiian," because it is similar to the volcanic eruptions you can see in Hawaii National Park today. The eruption begins with gas-charged lava "fountains" and then changes to basaltic lava flowing from fissures. The main differences between the basalt eruptions that occurred here and those in Hawaii are the size of the lava flows and volcanoes. The Hawaiian islands are huge volcanoes made up of many lava flows. The Ice Harbor volcanoes are very small, but the lava flows were some of the largest volcanic eruptions that have occurred on Earth.

The Monument: A Living Geologic Laboratory

The Hanford Reach National Monument's relatively undisturbed land with its native plants and wildlife is the pride of the Columbia Basin. Its pristine condition is part of the natural history of our land and something we all will be proud to pass down to future generations. Nature has been creating this Monument for millions of years. Preserved at the Monument is the geologic history of the Columbia Basin and geologic features found nowhere else. Here, at our fingertips is a detailed record of the last 17 million years of the Columbia Basin.

The Hanford Reach of the Columbia River is but the most recent chapter in the Monument. The Reach itself is only a little more than 10 thousand years old, although the last 7-million-year history of the river is preserved along its banks in the White Bluffs. At the end of the Pleistocene when the last glacial Lake Missoula broke free from its ice dam and drained out of the Columbia Basin through Wallula Gap, it left the Hanford Reach as its legacy for future generations, including Kennewick Man.

The oldest rocks on the Monument are located on 17-million-year-old Rattlesnake Mountain, which towers over the Fitzner/Eberhardt Arid Lands Ecology Reserve (ALE) unit. These are the rocks of the Columbia River basalt that erupted during the Miocene Era. As lava covered the land, the Pacific Northwest east of the Cascade Range was caught between forces in the Earth pushing from the north and south. These forces buckled the basalt into great arches that rose above the surrounding land. Rattlesnake Mountain, Yakima Ridge, and Umtanum Ridge are just three mountains on the Monument that were created then.

Exposed on the flanks of Rattlesnake Mountain are lavas that are part of the largest lava flows known on Earth and, for that matter, in the solar system. They preserve the history of our area between 10 and 15 million years ago when the Columbia Basin was subtropical and trees that were the early ancestors of the giant redwoods grew on the land.

Between the lava flows that make up the Monument are sediments deposited by the ancestral Columbia River system as it flowed to the Pacific Ocean. These sediments record how the river gradually changed its course in response to the growing mountains. The earliest sediments show a river flowing along the northwest boundary of the Monument before turning west to Yakima, then south to Goldendale. Younger sediments show how the ridges forced the river to flow south along the western part of the Monument and then southwest to Granger and Toppenish before turning south.

The gravels we see at the base of the White Bluffs and in wells at the U.S. Department of Energy Hanford Site tell of a river forced still farther east. The Columbia River, joined by a river flowing from the east that would eventually become the Snake River, meandered across the Monument and flowed first up the present Yakima River Valley, then later, through its present water gap, Wallula Gap.

The White Bluffs sediments tell how this river lazily meandered across the land, leaving gravels where the river flowed and sands and silts in its floodplain. Those sediments are a treasure trove of fossils, indicating a rich and diverse ecology. High on the cliffs are thick lake beds that also contain fossil remains of fish and plants that fell in the water.

The cliffs of the White Bluffs tell how 3 million years ago western North America started rising in elevation. As the land rose, the Columbia River began to cut down through all the sediments it had deposited across the Monument, leaving only the White Bluffs. This set the stage for the final act: the spectacular cataclysmic floods of the Pleistocene epoch that formed the spectacular Hanford Reach.

The Monument also includes one of the least known but truly unique features of the Columbia Basin: the active sand dunes across from Ringold Coulee. These spectacular dunes owe their existence to both the oldest and youngest geologic features of the Monument. The strong westerly winds that blow across Rattlesnake Mountain sweep down across the Hanford Site and out of the Monument through Ringold Coulee, which was created by the Ice-Age floods. The low notch made by Ringold Coulee in the White Bluffs directs these winds, which keep the sand dunes forever active and the Hanford Reach National Monument a living geologic laboratory.

If you wish to have a geologic map of the Monument, my colleague Karl Fecht and I published two maps in 1994 that encompass the area, the Richland and Priest Rapids 1:100,000 maps (OFR 94-8 and OFR 94-13). In addition, a map of the southeast quadrant of Washington (GM-45) that includes the Hanford Reach National Monument also was published. These maps are available through the Washington Division of Geology and Earth Resources in Olympia. The address is: WADGER, P.O. Box 47007, Olympia, Washington 98504-7007.

The Geology of Kennewick Man

One of the most interesting scientific stories of my career is Kennewick Man, a 9,000-year-old skeleton found in 1996 by several college students during the Columbia Cup races in Kennewick. The skeleton originally was thought to be a relatively recent homicide, but work by archaeologist Jim Chatters soon showed this was not a recent inhabitant of the basin.

The geologic story of Kennewick Man is just as fascinating as the archaeological story. Scientists Jim Chatters, Karl Fecht, and I have been unraveling the geologic history of the Columbia Basin for more years than we'd like to admit. Kennewick Man is helping fill in some holes in our story.

Kennewick Man was found in sediments at the bottom of the Columbia River near the current shoreline. We know these sediments were deposited either during the Ice Age or shortly after it. Four major floods and as many as 40 smaller floods were unleashed across eastern Washington when ice dams broke during the Pleistocene. These floods produced the Channeled Scablands of eastern Washington and left behind thick gravel deposits and terraces when floodwater rushed through our area. Fine-grained sediments were deposited when floodwaters backed up behind Wallula Gap as they tried to drain out of the Columbia Basin.

Deposits from two of the oldest glacial floods still remain a few miles to the west of Columbia Park, suggesting Kennewick Man rested on sediments as old as 200,000 years. The youngest of these humongous glacial floods occurred about 13,000 years ago. Deposits from this flood can be found all around the Tri-Cities up to an elevation of about 1,200 feet. The flood deposits are easy to identify, and their age is easy to determine because shortly after the flood, Mount St. Helens erupted, blanketing our area with two ash layers. The time of this eruption has been determined to be 13,000 years ago. These flood sediments originally filled the Columbia Valley to nearly 1,000 feet.

The 13,000-year-old glacial flood sediments are older than Kennewick Man. Places in the Tri-Cities where we find these sediments show that Kennewick Man came after the flood.

Following the 13,000-year-old glacial flood, at least three smaller floods occurred. These smaller, later floods eroded much of the 13,000-year-old flood sediments from along the main track of the Columbia River to its present level. They carved the floodplain that makes a prominent terrace in our area at about 400 feet elevation.

Columbia Park is part of this terrace, and Kennewick Man was covered by the sediments that make up this terrace. Thus, geology tells us Kennewick Man is older than this terrace.

Until the Kennewick Man research, we thought this terrace formed around 8,000 to 9,000 years ago, based on various pieces of evidence, including volcanic ash.

While U.S. Army Corps of Engineers geologists were examining sediments that make up this terrace, they noticed a layer of white material they thought looked like volcanic ash. After examining their sample, we confirmed it was ash and, based on its properties, identified it as ash from the 6,700-year-old eruption of Mount Mazama (now Crater Lake National Park). This ash layer confirmed that the geologic age of this terrace is older than 6,700 years. This ash find was truly fortuitous, because even though Mount Mazama ash can be found around the Tri-Cities, it usually isn't well preserved, and we rarely find it in place.

The geologic history of Kennewick Man nicely fits with the radiocarbon age determined for the skeleton. We now know that Kennewick Man lived in our area after the great floods of the Pleistocene, but he may have been around to see or get caught by one of the last floods.

We hope archeological studies will tell us if Kennewick Man lived at this site or died elsewhere and the Columbia River chose his final resting place.

White Bluffs

Not only do the White Bluffs provide us with one of the spectacular landmarks along the Hanford Reach, but as a bonus, they provide a glimpse of our past. After the last great basalt eruption occurred 10 million years ago, the region was left as a flat, featureless plain. Only the Columbia and Snake rivers broke up this monotonous landscape. These rivers seemed lost in a new world because the channels they had occupied before the eruption were buried under hundreds of feet of lava. After the last basalt eruption, the rivers had to start over doing what rivers do: cut channels and deposit sediment. Thus began the history of the White Bluffs.

For more than 7 million years after the lava wiped out their world, the Columbia and Snake rivers recorded the geologic events that shaped the region and gave us our present landscape. They were truly the first scribes of the "modern" Columbia Basin.

The rivers began their work in the basin by first settling into low areas. They were helped along by the growth of ridges like Rattlesnake Mountain, which allowed natural valleys to form between them. As the ridges grew, the valleys became more established and well defined.

With time, the Columbia and Snake rivers carried enough sediment to begin to fill the valleys. We find gravel where the main channel was and finer grained sediment where the old rivers flooded their banks. If you look at the White Bluffs now, you'll see this. Near the present Columbia River at Taylor Flats, you find gravels made up of all kinds of rocks. Above the gravel are finer silts and sands, which show the main channel moved away from the White Bluffs. At the top of the White Bluffs are lake beds that tell us that 3 million years ago extensive lakes covered much of the basin.

By the time the lakes were thriving at the White Bluffs, the entire area had filled with sediment to the level of the top of the White Bluffs. Geologists find these lake beds from the base of Badger Mountain to the base of Rattlesnake Mountain. Try to imagine our area as one large expansive floodplain of the Columbia and Snake rivers with its surface at the elevation of the White Bluffs. That landscape, 3 million years ago, was certainly different than the one we have now.

One question geologists have had for many years is—what happened? What caused the Columbia and Snake rivers to change from rivers that deposited sediment to ones that eroded sediment and formed the White Bluffs?

There wasn't a good answer for many years. We argued that something happened in the western United States that caused our region to begin to rise in elevation and force the rivers to cut into the sediment. That answer always seems to raise even more questions.

However, we might be closer to the answer thanks to a colleague at the University of Washington. One year I had the opportunity to take a field trip along the east side of the Cascades with Dr. Eric Cheney. Dr. Cheney argued that the most recent uplift of the Cascade Mountains began about 4 million years ago. Like all mountains, the Cascades go through growth spurts, but geologists had believed that the most recent phase of the Cascades began about 8 to 10 million years ago.

His argument for the age of the most recent uplift in the Cascades was based on sediment from rivers flowing from the Cascades. He found there was a significant increase in sediment from the Cascades starting about 4 million years ago. If a mountain range has been worn down by erosion, there won't be much sediment in its rivers. But if all of a sudden the mountains start to rise rapidly, then there will be a significant increase in sediment as the mountains grow.

An age of 4 million years for the most recent uplift along the Cascades is a new idea but it also provides a possible explanation to the mystery of the White Bluffs. It implies that as the Cascades grew, the area as far east as the White Bluffs began to rise with it. Our uplift was certainly not as great as the Cascades, but it was enough to change the landscape. Perhaps we should start thinking of the White Bluffs as the little sister of the Cascades.

Wallula Gap

You can't miss Wallula Gap as you drive south from Pasco along U.S. Highway 12. It stands as the last great Columbia River water gap in Washington and makes a fitting monument to this great river as it leaves the Columbia Basin. Lewis and Clark mention the gap in their journal and spent several days here in October 1805 on their way to the Pacific Ocean.

Because the Columbia River didn't begin flowing through the gap until less than 5 million years ago, it probably is the youngest of all the Columbia River water gaps in the state. As you look at Wallula Gap, you see the face of a fault, Wallula fault, along which the Horse Heaven Hills have risen nearly a half mile in the last 8 million years. The ridge still is rising along this fault, but it never was a match for the Columbia and its ability to erode.

You can see the Wallula fault at several places. On the east side of the gap, the fault is exposed just west of the Yelpit siding on the railroad tracks. The siding got its name from the great chief Yel-lep-pit, who met Lewis and Clark near Wallula Gap. The best place to see the fault, however, is near the junction of State Route 730 and U.S. Highway 12. The fault lies directly under the straight stretch of U.S. Highway 12 that comes east from Walla Walla toward the river and the continuation to State Route 730.

Coming from the Tri-Cities, if you turn right (west) onto State Route 730, then turn into the port area a few hundred feet farther on, you can see a series of basalt outcroppings on the east side of the gravel port road. There are four lava flows here: the oldest (14 million years old) is exposed in the railroad cut, and the youngest (8.5 million years old) is exposed near the grain elevators.

Only the oldest of these lava flows is present above the road on the Horse Heaven Hills. The ridge had grown too high by the time the younger flows were erupted and reached the gap. This tells us that the Horse Heaven Hills were becoming a major ridge by that time.

But what about the river? Actually, there was a river flowing through the gap when the younger lava flows were being erupted. When geologists examined the rocks at Wallula Gap, they found an old river channel filled with gravels and basalt.

The basalt in the channel is the youngest (8.5-million-year-old lava) flow in the port area on the west side. You can see the channel if you stand at Two Sisters Park, a few miles ahead on State Route 730, and look to the highest point on the cliff on the west side of the river.

This old river channel told us that Wallula Gap was a water gap for many millions of years, but the gravels are not from the Columbia River. The gravels are made of rock from the part of Idaho that is now drained by the Salmon and Clearwater rivers.

The Columbia River comes from Canada, and we know that 8 million years ago the Columbia River was flowing across the Hanford Site and into the Yakima River near Horn Rapids. But it wasn't flowing down the Yakima Valley toward Wallula Gap at this time; it was flowing up the Yakima Valley past Benton City and Prosser to Satus Pass on U.S. Highway 97!

The Columbia River didn't flow to Wallula Gap until the rising Horse Heaven Hills and Rattlesnake Mountain reversed the tilt of the valley from west to east, forcing the Columbia to Wallula Gap. There the Columbia joined the other river and captured the gap, staking its claim to the area.

Geohazards

Geohazards

A colleague and I were in the field the other day, and we started talking about geologic hazards. We got on the topic because we were driving through a long, narrow canyon that had sheer walls with occasional rock slides along the road.

When we think of geologic hazards in the Pacific Northwest, the first thing that usually comes to mind is an earthquake or a volcanic eruption. We live in "Cascadia" where large earthquakes are to be expected, and where volcanoes are major tourist attractions, but it's easy to overlook the more common hazards. Landslides and avalanches in the mountains are some of the most common hazards, but a boulder falling off a ledge like at Wallula Gap, which did happen, can be very dangerous.

Talk to any geologist and they'll be happy to tell you about the worst hazard they know. I guess we all rate hazards on how uncomfortable you get when you're at that hazard. When Mount St. Helens erupted in 1980, most geologists thought of the hazard that it presented as minor in comparison to the excitement of a real live volcanic eruption. It was a chance to see something that most of us only can read about in textbooks.

I guess I'm like most geologists and look at earthquakes and volcanic eruptions as "geology in action" and less so as hazards. The more common hazards are the ones that I take more seriously. The one area in the Pacific Northwest I rate as my Number-One Hazard is along U.S. Highway 12 near White Pass.

A few years back, the stretch of road from Rimrock Lake up to near Dog Lake, below White Pass, slid away. It was a great example of a landslide of unstable ground. It took the Washington Department of Transportation quite a few months to rebuild that part of the road and get it back to two lanes. They had to refill the landslide area and stabilize it with "engineered methods."

That stretch of the highway has everything you need to make it my number-one hazard. When I look at it, I wonder how long before Mother Nature will teach the engineers another lesson. It took nearly 200 million years to create that hazard, and nature continues to sharpen her skills on that one.

The valley that Rimrock Lake sits in is a classic glacial valley carved from the rock during the Ice Age. Glacial valleys are "U"-shaped; that is, the valley has a wide bottom and very steep walls. Highway12 is cut right into the side of the steep north wall of that glacial valley. The rock that makes up the valley is over 150 million years old and very weak. It was not hard work for the glacier to carve that rock.

However, the rock that sits above the old, weak rock is young and hard. It is the pink rock that you see when you leave the valley. That hard rock is lava from a young, but extinct, volcano that rises above the east side of Dog Lake. The volcano is Spiral Butte.

The Spiral Butte lava is part of the hazard. Although this lava is hard, it too has weaknesses: lots of joints and factures. There are so many joints and fractures that the rock looks like it's peeling off the mountain, and it is doing just that. The fractures are oriented all different directions, but one direction is particularly bad. It follows the hill slope so the rock has a natural slide plane.

Winter and spring are particularly bad times of the year on this part of White Pass. Water gets into these fracture planes and freezes. As the water freezes, it expands and loosens the rock. Given enough time, a rock slab can break free and slide down the fracture plane. Once I drove over White Pass and saw a truck sitting by the highway. Such a slab had crushed the cab.

One year, I watched as the highway department repaired the latest hazard along the road. The hazard was a large slab of rock at the last bend before Dog Lake. The slab was separated from the main rock by a vertical fracture in addition to the nearly horizontal joints. Every time I drove over White Pass, I watched the progress the Department of Transportation made as workers installed a chain mesh over the rock and attached it to the slab and the mountain with rock bolts to keep the rock from falling onto the roadway.

The combination of the older weaker rock in the lower part of the valley, which is prone to landslide, and the highly fractured, younger lava higher up made me not want to ignore the signs posted along that part of the highway that year: No Stopping or Standing. There weren't any signs saying the speed limit was strictly enforced there.

Cascade Volcanoes

On clear days in the Columbia Basin, Mount Adams and Mount Rainier stand out like white diamonds above the gray rocks of the Cascade Range. A colleague who flew from Seattle to the Tri-Cities recently commented on the spectacular view of the Cascade peaks from the plane. The nearly straight line of snow-capped peaks stretching from the Canadian border way down into Oregon impressed him. He wondered what the area would have looked like if he had been here a few million years ago.

The answer to his question is that it probably did not look too different from what it does now because the Cascade Range is a volcanic mountain belt. Born less than 40 million years ago, the Cascade Range is relatively young from a geologist's point of view. The high Cascade volcanoes that we now call by such familiar names as Mount Rainier and Mount Hood are all less than 500 thousand years old—mere babes in geologic time. These young volcanoes are built on the old, eroded shells of generation after generation of once majestic volcanoes, equally as beautiful as any of the present ones, and equally as dangerous.

Anyone in the Pacific Northwest born before the 1980 eruption of Mount St. Helens has tasted the power hiding behind those beautiful exteriors. Most of us think only in terms of how the volcanic hazards affected the west side of the state. Here on the east side, we think of volcanic hazards more as a nuisance from the ash. However, any city close to either side of the Cascade Range can be impacted by the volcanic eruptions.

Yakima is a good example of living in the shadow of the mountain. If you were to ask any resident of Yakima if the city should worry about volcanoes, they probably would say ash was the only hazard. That may be true for now, but maybe not for the future.

The prominent hills west of Yakima are a popular place to build houses. The hills offer a great view of the valley and the Cascade Range, but they are there because of a major volcanic eruption in the Cascades less than 1 million years ago.

The Goat Rocks Wilderness Area, about 45 miles west of Yakima, once was a major Cascade volcano. It erupted between 3.2 million years ago and 200 thousand years ago. At that time, the Goat Rocks looked pretty much like Mount Rainier, reaching

15,000 feet high during the peak of its activity. Now, all that remains of the original volcano is Black Thumb peak, which Ice-Age glaciers carved.

Although it's a long drive from Yakima to the Goat Rocks, the hills on the west side of Yakima owe their existence to that volcano. The hills are formed from the Tieton lava flow, which is less than 1 million years old. Geologists call the Tieton lava an andesite to describe its special characteristics. Andesite normally does not produce lava flows that go great distances. The lavas tend to form large volcanic cones that give the high Cascade volcanoes their distinctive shape.

The Tieton andesite is different. This lava flow is the longest andesite lava flow known in the world, worthy to be a Guinness Book of Records placeholder. As the Tieton lava erupted in the Goat Rocks, it flowed more than 50 miles eastward down the ancestral Tieton and Naches rivers to reach Yakima. In some places, the flow is more than 500 feet thick.

By the time this lava flow stopped, it covered much of the western part of Yakima. To geologists, this is an impressive distance and a reminder that the Cascade Range is there because of volcanoes.

The Goat Rocks volcano may no longer be active, but the road to White Pass is paved with the remains of many volcanoes. Some are as old as 35 million years and some are very young. The Cascade volcanoes familiar to us today most certainly will be replaced with new ones in the future.

Mount St. Helens

The eruption of Mount St. Helens, on May 18, 1980, was the largest Cascade eruption in recorded history. To geologists, it was a real education in the fury of a natural process that is just a common part of Pacific Northwest history. Although named by the British explorer George Vancouver for his friend, Lord St. Helens, Native Americans had a more appropriate name: Tah-one-lat-clah—Fire Mountain.

In geologic time, Mount St. Helens is a young Cascade volcano, probably the youngest of the chain. Its birth is recorded in lavas only 35,000 to 40,000 years old, compared to the nearly 500-thousand-year-old lavas of most other Cascade volcanoes like Mount Rainier.

From its birth, Mount St. Helens had an active life, with many explosions followed by rebuilding its cone with lava. The Mount St. Helens we knew before 1980 began about 400 B.C. By 1647 A.D., Mount St. Helens had built that impressive landmark that was on all the postcards. After that, however, the volcano went dormant for about 150 years.

Beginning around 1800, the volcano came back to life. Explosive eruptions were followed by lava eruptions that were followed by creation of a small lava dome near the summit of the volcano.

The last eruption before 1980 occurred in April 1857 when the Pacific Northwest was still sparsely inhabited. Reports of dense smoke and fire coming from the volcano appeared in newspapers west of the Cascades. Then, again, Mount St. Helens slipped back into a quiet sleep.

At 3:45 p.m. on March 20, 1980, Mount St. Helens abruptly signaled her reawakening with a magnitude 4 earthquake. This was followed by a series of earthquakes that peaked on March 25th. The earthquakes told geologists that magma was forcing its way up into the volcano after a 123-year rest.

The first eruption since 1857 occurred on March 27, 1980, at 12:36 p.m. when the rising magma caused groundwater to flash into steam. This opened a crater about 250 feet across at the summit and threw rock into the air. That was the beginning of a series of steam and ash eruptions, some of which saw clouds billow as high as 11,000 feet into the sky.

On March 29, a second crater opened on the summit of the volcano, and by April 8, the two craters had merged to form one large crater 1,700 feet across and 850 feet deep. By March 30, geologists had recorded as many as 93 steam and ash eruptions. The volcano was certainly coming out of its sleep.

Earthquakes continued as magma rose in the volcano and deformed it. By March 27, a series of fractures had cut the summit and north side. The fractures formed in response to magma forcing the surface to bulge out. By the end of April, the bulge had grown and had moved out to the north over 260 feet. By mid-May, the bulge was easy to see from a distance because it had moved out more than 400 feet.

The last week of April and first week of May were relatively quiet ones at Mount St. Helens. There were no steam or ash eruptions. But shortly after, the summit crater began almost continuous activity as the bulge continued to swell.

The morning of May 18 started out relatively quiet, but at 8:32 a.m., a magnitude 5.1 earthquake occurred directly below the north side of the volcano. At the same time, the mountain began to fall apart. The north side slid away, creating one of the largest landslides ever seen. Within seconds, gas and ash streamed out of the mountain into the sky and followed the sliding rock into the valley below. This was the beginning of a very long week in the Pacific Northwest and a major change in our perception of the Cascades.

Many people lost their lives that day because they didn't take the volcano seriously. A graduate student and friend working at the Hanford Site, Jim Fitzgerald, was willing to risk the dangers to learn from the volcano first hand. He paid the ultimate price for his education that day. Those of us who knew Jim always think of him on the anniversary of the eruption.

Now, years later, the Cascade volcanoes once again look peaceful. The peacefulness of the Cascade volcanoes is deceiving, however. The Pacific Northwest Seismic Network records earthquakes under the Cascade volcanoes all the time. These earthquakes remind us the volcanoes are not sleeping but merely resting.

Earthquake/Tsunami Preparedness

April is Earthquake and Tsunami Preparedness Month in Washington and Oregon. In eastern Washington, we highlight these natural hazards to remind citizens that we live in earthquake country and need to be prepared in the event of an earthquake.

Living east of the Cascade Range, it is easy to become complacent about earthquakes because we rarely feel any. However, in the last two decades, geologists have found considerable evidence that western Washington and Oregon experience earthquakes as large as magnitude 8 and 9 every 200-400 years. This is a sobering thought when you learn that the last big one was 200 years ago!

Eastern Washington and Oregon are not immune to large earthquakes. The large ridges around us like Rattlesnake Mountain provide the hard evidence for this. Earthquakes are the growing pains that accompany the ridges as they develop, and they're not yet done growing. Although you don't feel most of them, we experience earthquakes every year in the Columbia Basin.

Unfortunately, it doesn't take an "old timer" to remember when the Columbia Basin had a large earthquake. All we have to do is go back to July 15, 1936, when a 6.1-magnitude earthquake occurred in the Walla Walla area. This was followed by several aftershocks. The damage was described by the U.S. Geological Survey:

> Four miles west of Milton-Freewater the ground was cracked over an area of 1,200 feet to 1,500 feet long by 50 to 100 feet. ...One stucco house [in Umapine] was badly damaged and a concrete residence fell to the ground. High school buildings...were pulled apart. Water wells changed. Cracks appeared in the ground. Large cracks appeared 3 miles west of Umapine. At a ranch between Umapine and Freewater cracks appeared...and water was forced out of the ground in a dozen places.

Although most damage was confined to the valley floor, the exact fault that caused the earthquake has never been identified. It was probably one of two main faults that define the Walla Walla Valley. One is the Blue Mountains front east of Walla Walla, and the other marks the Horse Heaven Hills to the south.

Faults are numerous and not hard to find near Walla Walla. While trying to find the fault that caused the 1936 earthquake, geologists found many young faults that were not previously known.

An easy fault to find near Walla Walla is the Dry Creek fault about 6 miles south of Milton-Freewater on Oregon State Route 11. This fault is part of the Horse Heaven Hills fault system. The fault crosses the road a few miles before the turnoff to Sprout Springs Ski Area. Just as you drop into Dry Creek, there are some vertical fractures (faults) in rock outcroppings along the side of the road. These faults have cut through basalt and some overlying sediments. The sediments were deposited during the Ice Age.

The Dry Creek fault is too old to be the source of the 1936 earthquake. Sediments deposited in the past several thousand years cover the fault but are not cut by it. This is how geologists determine the age of a fault.

We use the amount of offset in rocks cut by faults and the lengths of the faults that moved to estimate the magnitude of an earthquake produced by a fault. The most recent movement on the Dry Creek fault (probably several thousand years ago) is only about 1.5 feet, which wouldn't produce a magnitude 6 earthquake.

The 1936 earthquake is not the only earthquake to have occurred near Walla Walla. The Eastern Washington Seismic Network operated by Pacific Northwest National Laboratory for the U.S. Department of Energy has recorded many other earthquakes around Walla Walla in recent years. In 1991 and in 1992, for example, earthquakes with magnitudes of 4.3 and 4.1 occurred there, and residents felt them. Another earthquake in 1999, magnitude 2.5, occurred near the estimated location of the 1936 earthquake.

Hopefully, earthquakes like the 1936 earthquake near Walla Walla won't occur again for some time. Unfortunately, we know that an earthquake that size will occur again, we just can't predict when. If we did have a large earthquake, would you be prepared?

The easiest place to find out how to prepare for an earthquake and what to do during an earthquake is in the Community Pages in a telephone book. April is a good time to make sure your family is prepared and knows what to do—just in case.

Rattlesnake Mountain Earthquakes

Here in the western United States, earthquakes are a part of life. Luckily, we don't experience many big ones in the Columbia Basin. Although in 2003, a magnitude 2.9 earthquake occurred near Ellensburg and a magnitude 3.2 near Yakima.

In 2002, we had more than 500 earthquakes in the Columbia Basin, and 42 in the local region surrounding the Hanford Site. Most were small earthquakes, which were less than magnitude 3 and, more commonly, less than magnitude 2. However, that doesn't mean we can't have bigger ones. It only means the big earthquakes come less often.

The main sources of large earthquakes in the Columbia Basin are ridges like Rattlesnake Mountain, the Horse Heaven Hills, and Badger Mountain. These ridges all have major fault zones along their bases. Faults occur where the rock has broken and moved. The larger faults are capable of producing earthquakes up to magnitude 7.5.

Most earthquakes we experience are either random events not specifically related to any known geologic feature like the faults along the ridges, or swarms consisting of anywhere from a few earthquakes to a lot of earthquakes in one small area. These swarms may last only a few days or they may be active on and off for years.

Some of the more active swarms near us are the Wooded Island swarm north of Richland, the Coyote Rapids swarm along the Hanford Reach east of Vernita Bridge, and the north side of the Saddle Mountains south of Royal City. These swarm areas have been active on and off for much of the time we have recorded earthquakes in the Columbia Basin.

Every once in a while we have geologic features where a lot of earthquakes occur in a small area over a short period of time. This happened in May and June 2003, when 16 earthquakes occurred on the south slope of Rattlesnake Mountain approximately 6 miles from the top of the ridge. All the earthquakes were small and had magnitudes less than 2. They were so small that no one could have felt them except for the earthquake detection instruments at the Pacific Northwest National Laboratory in Richland.

These earthquakes occurred deep in the Earth (about two to three miles), which placed them near the bottom of the basalt layer that covers our region. The basalt overlies sediments that were deposited by the ancestral Columbia River and its tributaries about 30 to 50 million years ago.

One of the most unusual things about this earthquake swarm is that it was centered over a small geologic feature that extends northwest of the Horse Heaven Hills between Benton City and Prosser. Geologists call this feature The Badlands anticline. An anticline is just a place where rock has been folded into an arch. The Badlands anticline gets its name because the great Ice-Age floodwaters eroded the rock at the top of the arch leaving a rugged landscape on the north side of the Yakima River.

We don't know why these earthquakes occur where they do or what caused them. That's the problem with earthquakes, they're unpredictable. We should be thankful for all the small earthquakes we see through the year, though. They tell us the pressure building up in the Earth is being released. Earthquake swarms such as on Rattlesnake Mountain are little reminders that the rock can't contain the pressure forever.

Alaska Quake Meets the Columbia Basin

The largest earthquake known to occur in the world in 2002 struck central Alaska on Sunday, November 3, at 2:12 Pacific Standard Time. The epicenter was located approximately 75 miles south of Fairbanks and 176 miles north of Anchorage. It caused countless landslides and road closures, but minimal structural damage, amazingly few injuries, and no deaths.

The earthquake resulted from a slip on the Denali Fault, a strike-slip fault, which stretches more than 500 miles across the state of Alaska and extends southeastward into Canada. A strike-slip fault is a fault like the San Andreas Fault in California that moves mainly horizontally, not up or down.

Geologists in Alaska measured the scarps caused by the faults and found that the maximum horizontal offset measured on the Denali Fault was 22 feet. This was determined by measuring the offset in the Tok Highway cutoff, a road that goes from Tok to Glenallen and intersects the Alaska Highway. The road crosses the fault at nearly a right angle, and the horizontal movement of the fault caused the road to literally be offset.

The earthquake was reported to have been felt in many parts of the lower 48 at many localities, even as far south as Louisiana. Because the Pacific Northwest is much closer to Alaska, it's not surprising that we had most of the reports. Some of the most interesting observations were in Lake Union in Seattle and Lake Chelan, which were reported to have sloshed around because of the earthquake.

The earthquake had a magnitude of 7.9, which ranks it up there as a really big one. Such a large earthquake will usually be picked up on seismometers around the world, but to feel the ground motion from one that far away is not typical. Pacific Northwest National Laboratory's Seismic Monitoring Laboratory's 41 seismometers from around eastern Washington easily picked up the earthquake, but to our surprise, several of our instruments that measure strong ground motion were triggered as well.

We use a special instrument called a strong motion accelerometer to record the amount of movement of the Earth as earthquake waves pass through an area. The instrument typically is not used to locate the source of an earthquake and does not measure the earthquake's magnitude. This instrument is designed only to "trigger" when the Earth's surface moves enough that people usually can feel it. When we checked our instruments on the Monday following the earthquake, we found that two of the five had triggered. Our seismometers first detected the initial motions at 2:17 p.m., five minutes after the earthquake struck Alaska, and two strong motion accelerometers triggered at 2:25 p.m., eight minutes later.

Our main seismic sensors triggered on the P wave (compression wave), which is the first wave to pass through the area. After the P wave came the S wave (shear wave). Neither of these triggered the strong motion accelerometers.

What ultimately triggered the strong motion accelerometers was another type of earthquake wave called a surface wave. These are long period waves, which means the peaks of the waves, like ripples in a lake, are about 15 seconds apart.

The movement of the ground recorded by these instruments is reported by convention as a percentage of the acceleration of gravity. To get an idea of what this means, start by dropping a rock from your shoulder height. That rock will move from a standstill to the ground with an acceleration of 32 feet per second, every second. That is, it moves faster and faster toward the ground with a measurable increase in speed.

When an earthquake wave passes through an area, the movement of the ground changes from a near standstill into a cyclic motion. In other words, it accelerates. Most of the time, the movement is so small that only our extremely sensitive seismometers can detect it.

The Pacific Northwest National Laboratory's strong motion accelerometers recorded maximum ground movements of 0.15% of the acceleration of gravity. To put this in perspective, think in terms of your rock accelerating toward the ground at 32 feet per second, every second. The ground in the Columbia Basin moved only 0.15% of that. That isn't very much, but it came all the way from Alaska, and that is impressive.

Tsunami Warning Network

Since the disastrous Indonesian tsunami of December 26, 2004, the world has begun to realize the importance of real-time warning systems for natural hazards. The populations along the coast now understand they are especially vulnerable to the consequences of earthquakes that occur in oceans far away.

Every day, earthquakes occur in rock at the bottom of the oceans. A quick look at the U.S. Geological Survey's Earthquake Hazard's Webpage (http://earthquake.usgs.gov/) and you can see just how common they are. Luckily, the circumstances that produced the December 26 tsunami are not everyday events. On December 26, the movement of rock vertically along a plate-boundary fault combined with the length of the fault that moved provided perfect conditions for a major tsunami.

Geologists and seismologists in the United States have long known about the potential for large tsunamis. Our own Plate Tectonic boundary, the Cascadia subduction zone, has produced a long and impressive geologic record of earthquakes. A major branch of geologic research involves studying sediment deposits along the coastline that resulted from large floods and earthquakes. These types of studies in the Pacific Northwest have shown that earthquakes from the Cascadia subduction zone have produced devastation along the Washington and Oregon shores similar to that from the December tsunami in the Indian Ocean.

As long ago as 1965, an international agreement started the process of developing the present tsunami warning network. Currently, 26 nations are members of this system but, as many of you may have read in the newspaper, the main network is in the developed countries.

As part of this network in the United States, the National Oceanic and Atmospheric Administration operates two tsunami warning centers—one in Alaska and one in Hawaii. The Alaska center serves the Pacific Northwest. The objective of the tsunami warning centers is to detect, locate, and determine the size of an earthquake in the ocean that could produce a tsunami. If the location and magnitude of an earthquake meet certain criteria, then a warning goes out to coastal areas that could be impacted.

The earthquake network used to locate and measure potential tsunami-generating earthquakes is not too different from the earthquake monitoring stations operated by Pacific Northwest National Laboratory. The devices are designed to measure how much the ground moves and when earthquake waves pass a station. Many sites are needed to accurately locate an earthquake, and the sites must surround the location of the earthquake to accurately locate it.

In an area like the Pacific Ocean basin, setting up a network to locate and measure earthquakes is a daunting task. Just to detect and locate earthquakes at the Hanford Site, we need many monitoring stations spread throughout eastern Washington. The size of the task to monitor the Pacific Ocean explains why much of the area around the December tsunami didn't know until too late.

One of the big shocks to people of the Tri-Cites and surrounding areas is that we have one of the stations that is part of the tsunami warning network located here. The Hanford tsunami site is located in an old Nike missile silo on the Hanford Reach National Monument just north of Rattlesnake Mountain. The instruments were installed about 5 years ago as part of tsunami earthquake monitoring network by the U.S. Geological Survey in cooperation with Pacific Northwest National Laboratory. The U.S. Geological Survey selected Hanford because of the resident seismic staff at the laboratory and because Hanford is a seismically quiet area that allows the detection of earthquakes from far away.

Even though we have little chance of a tsunami coming up the Columbia River, far inland sites like Hanford are necessary to give the stability, sensitivity, and coverage needed to detect earthquakes halfway around the world. You might think of our corner of the desert as part of the first line of defense against tsunamis.

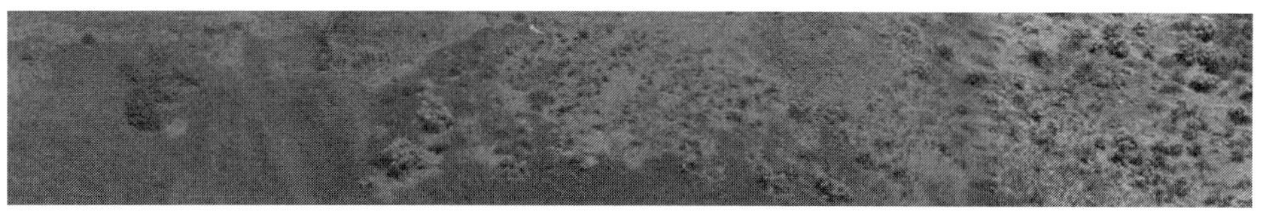

Around the Region

The Odessa Ring Craters

Columbia River basalt flood lavas are one of the great mysteries of the world. One fascinating feature of Columbia River basalt that has puzzled geologists is the ring craters of Odessa, Washington. Nestled in a 60-square-mile area in the valleys that surround Odessa are more than 100 circular craters.

The ring craters are found in the Roza lava flow, a 14.5-million-year-old lava flow that covered more than 15,000 square miles of the Columbia Basin. Geologists have examined the Roza flow all over eastern Washington, but only at Odessa are the curious circular craters found.

A typical ring crater is a circular-to-elliptical depression with a series of discontinuous concentric ridges. The craters range in diameter from several hundred feet to as much as 2,000 feet. Some have a central mound, and all probably were once filled with lava. The great Ice-Age floods that formed the valleys around Odessa removed the overlying soil and scoured out the craters.

Most ridges that make up the ring craters are Roza basalt intruded by dikes. Dikes are usually found only where magma has forced its way up to the surface from the Earth's interior. The Roza lava, however, erupted farther east, so these dikes cannot be the source for the Roza lavas.

The Odessa ring craters have been compared to explosion craters because of the concentric fractures and central mounds. But these craters formed when the Roza lava erupted. The evidence for this is in the dikes or, in this case, rootless dikes as geologists call them. The rootless dikes are Roza lava intruding Roza lava flow. This means that the lava was beginning to solidify as the craters formed. Something caused the crust of the lava to fracture and break and force still-molten lava up through the fractures in the crust of the lava. These rootless dikes originate not in the Earth's interior but in the Roza lava flow; thus, they have no roots. You can recognize these rootless dikes by the horizontal fractures and joints (like a stack of wood) that stand out in stark contrast to the vertical joints in the normal Roza lava.

What could have caused the ring fractures? We know they aren't volcanoes, but they look similar. A popular hypothesis among many geologists is that the lava flow encountered water, either in lakes or in an old river. If the Roza lava flowed into water, we would expect to see something called pillow lavas that are typical of Hawaiian lava flows. Pillows are just like the name describes: basalt shaped like the pillows on a bed. When basalt lava flows into water, the hot molten lava freezes instantly in pillow shapes, but the center of the pillow is molten. The molten center will break through the frozen crust of the pillow and form more pillows. We just don't see pillow lavas at Odessa.

Another hypothesis closer to fitting the rocks we see is that the lava flowed over shallow water or water saturated soil. The lava would cause the water to flash to steam, producing an instant explosion. This type of explosion would produce features similar to what we see at Odessa. This seems to provide the best explanation.

If you want to spend a nice day walking around some unusual geologic features, go to Odessa and explore these craters. The Bureau of Land Management (BLM) has set aside large areas of the land where the craters are found so you can hike and explore these features. In conjunction with the Odessa Chamber of Commerce, the BLM also has produced a beautiful, full-color guide to the ring craters and the surrounding area. This detailed guide has pictures of the craters as well as maps, road logs, and descriptions of many other fascinating geologic and scenic features that make the area around Odessa truly an area worth visiting.

Odessa is north of the Tri-Cities. To get there from Interstate 90, take State Route 21. You can get more information on the ring craters from the Odessa Economic Development Committee, Odessa Chamber of Commerce, P.O. Box 565, Odessa, WA 99159.

Pinnacles State Park

Pinnacles State Park near Wenatchee is not well known to most people in the central Columbia Basin, but rock-climbing enthusiasts know it well. This park is just east of Cashmere along U.S. Highway 2 and U.S. Highway 97 between Leavenworth and Wenatchee. The park is named after majestic towering spines of white rock that rise hundreds of feet above the Wenatchee River.

The rock that makes up the pinnacles is nearly pure quartz sandstone that was deposited by rivers between 45 and 39 million years ago. When you walk up and look at the rock, you can see all the features you would expect to find in modern-day river sand: channels, sand bars, and evidence of old shorelines.

However, there's a small problem with the rock that forms the pinnacles. It appears to be standing on end. The beautiful white pinnacles are old riverbeds that were tilted and now are slowly weathering away.

The history of Pinnacles Park is the history of the Wenatchee Valley itself. If we were to go back nearly 50 million years to the geologic epoch known as the Eocene, the Wenatchee area would look much like a tropical shoreline with rivers flowing into the Pacific Ocean.

About 45 million years ago, this peaceful setting began to change. A large depression began to form where the Wenatchee Valley is today. This depression more than 10 miles wide and oriented northwest much like the Wenatchee Valley is today.

The depression was marked by two major faults: one along the present north side of the valley and the other paralleling the south wall of the valley. The depression is called the Chiwaukum graben. Geologists use the word graben to describe an area like this that is bounded by faults.

As the graben floor sunk, rivers and streams carried sand from the surrounding highlands into the valley. The rivers brought in so much sand that the valley never looked deep. However, geologists estimate that more than 18,000 feet of sediments were deposited in the graben.

The valley subsided for about 6 million years. Then, 40 million years ago, conditions once again changed. This time, the change was the beginning of the Cascade Range. As the Cascades began to grow, the Wenatchee Valley was caught up in the mountain building. The faults that defined the edges of the graben once again started to move.

As the faults moved, the surrounding highlands rose with the Cascades. This time, however, the Wenatchee Valley was squeezed from the north and south as well. The sediments laid down by rivers millions of years earlier were now folded and faulted.

The river sands in contact with the two big faults fared the worst. Not only did they get folded, but they also were dragged along by the faults and turned on end. Pinnacles State Park lies along the northern fault. The spectacular pinnacles that give the park its name are those sandstones that were folded and dragged by the fault.

The last part of the story of the park is also the story of the present Wenatchee Valley. The highlands surrounding the Wenatchee Valley are outside the boundaries of the Chiwaukum graben. These highlands are there because they are made up of very hard rock. Just drive north along U.S. Highway 97 to Lake Chelan and you can see this first hand. These rocks make formidable cliffs.

The rocks of the Wenatchee Valley are the Eocene river sands. They are weak rock and easily eroded. The Wenatchee River was faced with a tough decision: try and erode its valley in the hard rock or the weak rock. You can see the choice it made. As the Wenatchee River slowly cut down through the Eocene sands, it uncovered the pinnacles and sculpted them into one of nature's works of art.

The Blue Lake Rhinoceros

Eastern Washington is world famous for the petrified wood found between Columbia River basalt lava flows. Ginkgo Petrified Forest State Park is a showcase for petrified wood, but it is just one of many localities where it is found in the Columbia Basin.

Rarely does one hear of animal fossils being found in the basalt. Clarkia Lake near the Washington-Idaho border is famous for Miocene Epoch insects and leaf fossils preserved in lake sediments along the basalt margin, but mammals have yet to be found there.

The best way to explain the plant and insect fossils and scarce animal fossils is that the animals were able to outrun the lava. Although the lavas could flow from the Washington-Idaho border to as far west as Portland in a couple of weeks, animals roaming the Columbia Basin could out distance the lavas with a brisk walk.

One of the few animals ever found in the basalt was near Soap Lake. It is affectionately called the Blue Lake rhinoceros. The Blue Lake rhinoceros was found in 1935 in Jasper Canyon just east of Blue Lake by people hunting petrified wood. The petrified wood hunters found a small cave about six feet long that contained a jawbone and a few other bone fragments.

Paleontologists made a plaster cast of the cave and found that the cave was shaped like a rhinoceros that had been encased in the basalt. The rhinoceros is called *Diceratheriom anneciens*, and it is 14.5 million years old (middle Miocene Age).

Why did the lava catch this animal and not others? Was it sick and couldn't outrun the lava flow? To answer these questions, we need to look at how the fossil rhinoceros was preserved.

The rhinoceros was found in the lower part of the Priest Rapids lava. A thin, sandy sedimentary deposit lies between the Priest Rapids lava and the underlying Roza lava. This layer is well known for petrified wood in the Columbia Basin. In the Saddle Mountains just north of the Tri-Cities, all types of trees have been found in this layer. At one locality, the Priest Rapids lava encases a sugar maple more than three feet in diameter.

The rhinoceros and the trees found in this layer indicate the climate was much warmer than today's climate. They show that, in the Miocene, the Columbia Basin didn't have long protracted winters as it does today. In fact, the worldwide climate record indicates that the Miocene was the warmest it has been in the last 40 million years.

A rhinoceros might seem like an unusual animal to find in the Pacific Northwest, but other North American fossil localities of the same age show it was common. The Miocene Epoch is called the Golden Age of Mammals. Life in our region then was very different than it is today. If you had been here in the Miocene, you also would have seen animals like the camel, some as tall as a giraffe, giant pigs as tall as an ox with skulls four feet long, and the infamous saber-tooth tiger. None of these fossils have ever been found in Columbia River basalts, but they have been found in the Ringold Formation at White Bluffs.

The way the Blue Lake rhinoceros was found in the lava tells a lot about how it got there. The Priest Rapids lava that encases the rhinoceros is pillow basalt. The pillow basalt indicates the Blue Lake area was probably a swamp or floodplain. The rhinoceros was not found in a standing or running position. Because of this, and that it should have been easy for the rhinoceros to outrun the lava, paleontologists think the animal was already dead when it was covered by lava. They believe it was bloated and floating in the water when the lava engulfed it.

Although the site of the Blue Lake rhinoceros is just south of Sun Lakes State Park, it is surrounded by private land. The easiest way to reach it is by crossing Blue Lake in a boat. A good landmark for Jasper Canyon is the junction of the road to Sun Lakes State Park and State Route 17. Jasper Canyon is due east, and the rhinoceros is located on the south side of the canyon.

Soap Lake

One of the most interesting aspects of living in a desert is what the limited rainfall does to the local geology. Besides not having a lot of erosion caused by rain, we also don't have many natural lakes. The natural lakes we do have are not the typical lakes you would find in wetter climates like the Midwest. Many of our lakes are better known for minerals than fishing.

Soap Lake, south of Dry Falls Park, is a good example of a natural lake in the Columbia Basin. Soap Lake formed during the Pleistocene Epoch and is part of the story of Dry Falls Park. When the floodwaters released from Glacial Lake Missoula reached the present site of Grand Coulee Dam, they turned south as they ran up against a small north-south fold in the basalt. As the floodwaters moved south, they began carving into the basalt and forming a canyon or coulee.

The resulting coulee now is called Lenore Coulee. Dry Falls Park sits at the north end of Lenore Coulee, and Soap Lake sits at the south end. Soap Lake and all the lakes south of Dry Falls Park formed when water filled low areas in the scabland topography that was carved by Ice-Age floodwaters.

Although Soap Lake is part of the Dry Falls story, it has its own geologic story separate from the floods. The story of Soap Lake centers around water. Soap Lake has a salinity or salt content that is just a little greater than seawater.

Soap Lake covers more than 800 acres, making it the third largest saline lake in Washington. It derived its name from the conspicuous froth that forms along the lakeshore, especially on a windy day. The froth looks very similar to soap suds in a bathtub or washing machine. The froth stays around for some time, so it has been called a tough foam; a light foam will disappear shortly after it forms.

At the beginning of the 1900s, Soap Lake was known as Sanitarium Lake, and its waters were believed to be therapeutic. As a result, it became a resort area and still enjoys part of that reputation today. The therapeutic properties of the water come from the dissolved salts that give it its saline character.

No freshwater fish can live in Soap Lake because it's too salty, but plants and other animals such as algae and small microscopic creatures called diatomaceae live there. If you were to look at low areas along the shoreline of Soap Lake when the water is low, you would find patches of white crystals where dissolved salts have formed crystals as the water dried up. Although the water is saline, and the concentration of salts is too low to naturally form crystals in the lake, patches of water will leave behind delicate crystals of salt as the water evaporates.

The crystals you see along the shoreline are made from salts of mainly sodium. The primary minerals you see are natron (sodium carbonate), mirabilite (sodium sulfate), and halite (sodium chloride) or common table salt. It has been estimated that about 628,000 tons of sodium carbonate, 300,000 tons of sodium sulfate, and 240,000 tons of sodium chloride are dissolved in Soap Lake.

So where did all these salts come from? We know Soap Lake isn't an old arm of the ocean because it formed as a result of the Missoula floods about 10,000 years ago. The answer comes from our desert climate. If you look at the salinity of all the lakes between Soap Lake and Dry Falls, you find they become more saline the closer you get to Soap Lake. The water that fills these lakes comes from groundwater and from the little rain and snow we have.

Each lake between Soap Lake and Dry Falls drains toward Soap Lake. As water leaves each lake, it carries dissolved salts derived from local rocks and soil into the next lake. However, Soap Lake is at the end of the line and collects the salts from all the upstream lakes. The water flowing into Soap Lake has nowhere else to drain, so the only way for water to leave Soap Lake is by evaporation, which leaves the dissolved salts behind. As time goes by, Soap Lake will continue to get more and more saline. Given enough time, Soap Lake will be another Salt Lake.

The Gorge Near George, Washington

If you have ever had a chance to see any of the various "big name" entertainers at the Gorge Amphitheater between George and Quincy, you probably were impressed by the tremendous view. The owners took advantage of a remarkable natural setting to enhance the performances.

The amphitheater is perched 800 feet above the Columbia River on the east side of a narrow gorge that extends more than 20 miles from Vantage to the south to Crescent Bar and Lynch Coulee to the north. Naneum Ridge forms the backdrop. It enters stage right in the Cascades near Wenatchee and exits stage left, west of the Tri-Cities.

The amphitheater where you watch the performance is built in a narrow stream tributary of the Columbia, but is dwarfed by the much larger amphitheater created by the forces of nature that carved the gorge and caused the ridge to rise. The stage uses the stars as its proscenium arch.

The geologic tale told at The Gorge begins more than 16 million years ago with the eruption of the basalt lava flows that cover most of eastern Washington. These lava flows extend one and one-half miles below the amphitheater, giving a good, sound foundation for the stage.

Even as the basalt lavas were cooling and hardening, Naneum Ridge was growing. As Naneum Ridge grew, it abducted the Columbia River and forced it to follow its present path from Wenatchee to near Priest Rapids Dam. The Columbia River just complacently settled into a low area formed by Naneum Ridge and the gentle slope of the land from Pullman and Spokane to The Gorge.

A battle between the basalt lava and the Columbia River took place until about 14 million years ago, when the last of the great basaltic lava flows to reach the area cooled, and the Columbia was able to conquer and erode through the pile of basalt.

This was a time of great change for the rivers of western North America. Geologists think the entire western part of our country underwent a time of regional uplift. That is, land elevations rose with respect to sea level as far east as Wyoming. The Columbia River responded by beginning a period of rapid downcutting. This is the same time that the Columbia began cutting the White Bluffs north of the Tri-Cities.

The story doesn't end there, however. The last great act was yet to play—probably one of the most spectacular of all the geologic acts of the Northwest. This was the cataclysmic floods of the Ice Age. Many times in a run of a million and a half years, ice dams in northern Washington broke, unleashing great torrents of water that cut quickly and deeply into the basalt and nearly filled the gorge. These are the floods that formed spectacular Frenchman Springs Coulee to the south and Lynch Coulee to the north.

Mother Nature, the author of our story, has yet to reveal her ending. The river still cuts, and the land also rises, and this play will undoubtedly outlive all the players at The Gorge.

Ginkgo Petrified Forest State Park

Although the Columbia Basin is now a semi-arid shrub-steppe, this area once supported a wide variety of trees that grew in a warm, temperate climate similar to that of the southeastern United States. That ancient forest is known as the Russell Forest, after pioneering geologist I.C. Russell. The forest flourished between 16 and 8 million years ago during the Miocene and is one of the largest petrified forests in the world.

The fossil record of this once large and varied forest can be found throughout the Columbia Basin, but is best preserved at the Ginkgo Petrified Forest State Park near Vantage. More than 50 genera and 200 species of trees are found in Russell Forest. Many of these trees are now extinct in North America but can be found in Asia, Africa, and Australia. One example of an extinct tree is the park's namesake: ginkgo, which was only recently reintroduced to North America. Other trees from the forest are now found only in eastern North America, including cyprus, elm, hickory, and honey locust. But some trees from the Russell Forest—fir, pines, oak, walnut, and sugar maple—still have living counterparts in western North America.

Today, all these trees occupy diverse climatic niches ranging from subalpine to warm, temperate climates. Why trees from such different growing conditions are found together is one of the mysteries of the forest.

The fossilized trees of the Vantage area are found in the Ginkgo flow, a 15-million-year-old lava flow of the Columbia River basalt. Most trees are found lying horizontally between pillows in the lava that formed when the molten lava came in contact with water. From their distribution, we know the trees must have been floating in a lake. Why a lake was there is another mystery.

A few years ago, some colleagues and I found a lahar near Ginkgo State Park that was deposited when the Russell Forest was growing. Lahar is the geologic term for a mud flow like the mud flows that flowed down the Toutle River when Mount St. Helens erupted in 1980. They form when a volcano erupts and its snow cap melts and mixes with ash, dirt, and rock. Lahars flow down the side of the volcano, carrying trees and anything else in their path.

We surmised that a volcano in the Cascade Range erupted and produced a lahar that flowed down its east flank into the ancestral Columbia River. Eventually, this mud flow got to the Vantage area where we think it blocked Sentinel Gap, the water gap through the Saddle Mountains south of Vantage. This dammed the Columbia River, forming the lake at Vantage. Trees carried by the mud flow from the Cascades floated in the lake like the trees at Spirit Lake when Mount St. Helens erupted.

Rapid burial of the trees by the Ginkgo lava was the key to their preservation and petrification. As the Cascade volcano was erupting, the Ginkgo lava also was erupting and flowing toward Vantage. The Ginkgo flow arrived last and flowed into the lake, burying the trees in a thick sheet of basaltic lava. The water chilled the lava and protected the trees and logs from burning.

The Ginkgo lava also heated the water surrounding the trees and helped accelerate the process of petrifying the wood. Silica that could more easily dissolve in the hot water replaced the wood and preserved the delicate wood structure. What now remains is the form of a tree but composed almost entirely of silica.

Leavenworth, Washington – The Christmas Town

No matter what the season, there is one place in Washington that reminds us of Christmas: Leavenworth. Leavenworth is a small town that lies deep in the Wenatchee Valley in one of the most dramatic landscapes in the state. The mountains rise rapidly from the edge of town and make you feel as if you're in the Bavarian Alps. This setting led the town to call itself the Bavarian Village, but in December, it would be better called the Christmas Village. The town transforms itself into a Christmas wonderland.

Leavenworth lies on the east side of the Cascade Range in the Chiwaukum graben. The word graben comes from the German word for grave, but the name is a simple way of describing the geometry of the rock; Chiwaukum is a local geographic name. The Chiwaukum graben is a large fault-bounded valley with the rocks on both sides being lifted higher than the rocks in the valley. If you have ever been to Leavenworth, you don't have to be a geologist to know that.

The faults that define the graben are called the Leavenworth fault and the Entiat fault. Both faults trend northwest, which is the same direction as the valley. The Entiat fault is on the north side of the valley, and the Leavenworth fault is on the south side.

The Leavenworth fault runs nearly through town, and it brings up rocks that are part of Mount Stuart. Mount Stuart is granite and is the 95-million-year-old root of ancient volcanoes. Give Mount St. Helens another 95 million years, and the roots of that volcano will look like Mount Stuart. The hard granite rock of Mount Stuart is one reason the mountains rise so abruptly above Leavenworth. The Alpine Lakes Wilderness is just above Leavenworth in this spectacular terrain.

North of the Entiat fault is hard rock that has been metamorphosed by the mountain-building process. These rocks are particularly well exposed along the Columbia River north of Wenatchee. Many of these rocks are much older than those of Mount Stuart, but they were sediments and volcanoes that originated a great distance away from Washington. They were added on to the state by the process of Plate Tectonics when the North American plate collided with oceanic plates about 100 million years ago. This process of accretion is how much of western North America came into being.

The rock that underlies Leavenworth is another story entirely. It's composed of sand and gravel that formed about 50 million years ago. These sediments are a prominent part of the Wenatchee Valley and are the main rocks you see along U.S. Highway 2. They formed by erosion of the older surrounding rocks.

The Chiwaukum graben probably started forming about 50 million years ago. Uplift of the Mount Stuart rocks and the old accreted terrane rocks helped provide the source for much of the sandstone at Leavenworth, but studies show much of the sands and gravels came from the northeast.

As the rocks of Mount Stuart and the accreted terranes rose, the valley began to sink and accumulate the sediment. It has been estimated that more than one mile of this sediment exists beneath the valley. Sediment from the eroding highlands still is accumulating in the valley.

The last part of the history of Leavenworth fits nicely with its Christmas setting. During the Ice Age, mountain glaciers moved south down the valley many times. Each time the glaciers advanced they carried sediments, and each time they retreated they left the sediments behind as a thin coating on the valley.

These glaciers helped shape the valley as well. The hard rock along the walls of the valley was smoothed a bit, and the softer rock was carved out, giving the valley its U-shape, characteristic of glacial valleys.

If you have a chance to visit Leavenworth, be sure to enjoy the scenery as well as the town. Mother Nature did a lot of work to add to your experience there.

Kittitas Valley

The Ellensburg area marks a transition between the Cascade Mountains and dry lands of eastern Washington. This is the Kittitas Valley where the Yakima River provides a clear, fresh source of water that has turned the basin emerald green. The geology is a wonderful mix of both the Cascades and the Columbia Basin.

The Kittitas Valley is sharply defined by the Frenchman Hills to the north, Manastash Ridge to the south, and the foothills of the Cascades to the west. Traveling east from Ellensburg, you gently climb to Ryegrass summit before dropping into the Columbia River Valley near Vantage. The Frenchman Hills, Manastash Ridge, and the Ryegrass summit all are the result of the Columbia River basalt lavas being folded and faulted by compressional forces in the Earth.

The nature of these ridges is as clear as a spring day when you drive into the valley on Interstate 82 from Yakima. After you cross the summit of Manastash Ridge, you get a beautiful view of the Kittitas Valley more than 1,500 feet below.

On a clear day from the summit of Manastash Ridge, you can see more different kinds of geology than from most places in the Pacific Northwest. On the horizon north of the Kittitas Valley, frozen magma that once fed volcanoes 90 million years ago now stands as the majestic Stuart Range. To the west, along the east flank of the Cascades, you'll see exposed rocks that extend for miles. Near Roslyn and Cle Elem at the west end of the valley are coal beds intertwined with 50-million-year-old river sands that came from Idaho. These coal deposits powered the Mid-Columbia in the early part of the last century, and for awhile, made backdrops for television shows.

Older than the coal beds but higher up on the ridges north of Elllensburg are lava flows. These old lavas hold a rare treasure: Ellensburg blue agates. Where these lavas have eroded from the ridges and their debris worked its way ever farther down into the valley, gravel deposits begrudgingly give up these prized gems to those who have the patience to look for them.

As the Columbia River basalts were erupting and the ridges were growing around Kittitas Valley, the Cascade volcanoes also were contributing to the geology of the valley. Ash and mudflows were pouring into the valley from early Cascade volcanoes, and when they weren't filling the valley with debris, streams and rivers slowly were carrying debris from the older Cascade peaks into the valley. You can see these deposits along the Yakima River if you drive the old highway from Ellensburg to Cle Elem.

Although the Kittitas Valley looks quiet, looks can be deceiving. The floodplain of the Yakima River is about 1,600 feet above sea level, but nearly one mile of sediment has filled in the basin on top of the basalt. To give this a little perspective, if you look at just one of the Columbia River basalt lavas at the top of Manastash Ridge, you would have to descend nearly a mile to find that same lava flow. That means that the same lava flow on Manastash Ridge is 3,200 feet below mean sea level in the valley.

To a geologist, this means that the bottom fell out of the Kittitas Valley starting about 15 million years ago when the basalts were erupting. The question is: Why did the Kittitas Valley subside when all the ridges were rising around it and the Cascade Range was growing? We may never know the answer.

One of the nicer things about geology is that the really hard questions need a lot of thought, and the best way to come up with answers is to find a beautiful viewpoint overlooking your problem and meditate. Sometimes the answers will come if you lean back, close your eyes, and let the sun help you think. If the answers don't come, so what! The nice thing about nature is you don't have to have an answer to enjoy a question.

Yakima River Canyon

If you have ever driven to Ellensburg through the Yakima River canyon on Canyon Road, I'm sure it made a lasting impression. This road was the main route between Yakima and Ellensburg before Interstate 82 replaced it, but it must still be one of the prettiest drives in eastern Washington.

The massive rocks that rise from the Wenas Valley to form Selah Butte mark the gateway. From that point, you follow the winding Yakima River where it cuts down through over 2,000 feet of Columbia River basaltic lavas to emerge just a few miles south of Ellensburg.

All along this route you see the lava flows rise then fall, only to rise again. The river cuts right though the heart of Selah Butte, North and South Umtanum Ridges, and Manastash Ridge, giving you a glimpse of the very roots of these mountains.

Each ridge is a series of folded rock that forms geologic features called anticlines and synclines. Ridges like Manastash Ridge are anticlines where the rocks are folded into up-pointing arches. Between the anticlines are synclinal valleys where the rock is folded into down-pointing arches. In each valley between adjoining ridges is a fault where the rock broke as it was folded into these spectacular mountains.

The Canyon Road route is certainly not a fast one to drive. It follows the river's course, snaking its way through the narrow steep-walled valley. In places, the road actually doubles back on itself as it heads north and then southeast before going north again.

When geologists look at the river pattern, it appears to be a mix of patterns. The narrow valley with the steep walls is characteristic of a "youthful" stream. Youthful streams have spent most of their time and energy doing nothing but cutting down into the Earth. They usually are straight and found in young mountains where they are rapidly carrying away as much rock as they can.

Then there is the winding pattern of the river through the canyon. This is a characteristic of an "old-age" river. When a river is said to reach old age, it no longer spends all its time cutting down into the Earth but begins to meander across the landscape. In its old age, it has done almost all the downcutting it can and is ready to kick back and relax a bit. If you look at the pattern of the Yakima River in the Ellensburg area, and from Yakima to Prosser, you'll see this same meandering pattern of the river. In these places, the river forms large loops or oxbows, as they are called, that look like someone was pushing a rope. These old-age rivers can have meandering channels that are 15 river miles long that cover only 10 miles in a straight line.

The question then is: Is the Yakima River in the canyon an old-age river or a young river? The answer is both. And here's why: The history of our region is a dynamic one. Sometime after the last of the huge Columbia River basaltic lavas flowed across the land and filled in all the low areas, the landscape was left as a flat, featureless plain. The Yakima River was presented with the perfect surface to form an old-age river. As a result, it began to develop a meandering pattern as it flowed from Ellensburg to the lower Yakima Valley.

However, soon after, the ridges began to grow in the canyon, the Yakima River was forced into action. It immediately had to start cutting into those ridges or it would be forced to find a new route to the ocean. The result was the old-age pattern superimposed on a youthful landscape.

All this geologic action has left a marvelous landscape in Yakima Canyon that you don't have to be a geologist to enjoy. All that you need is a nice afternoon.

Fife's Peak Volcano

The Cascade Range is one of the youngest mountain ranges in the western United States. It began to form less than 40 million years ago, which is young by geologic standards. However, as young as it is, geologists recognize two distinct periods of mountain building.

The first mountain building phase began about 37 million years ago with the earliest volcanoes and ended around the time the Columbia River basalt began to erupt 17 million years ago. The second period of mountain building began about 10 million years ago and is continuing with the volcanoes we see today.

The early Cascades Mountains probably didn't look much different than the modern range. Geologists who have been studying the Cascades have found traces of the early volcanoes. When they try to reconstruct their original size, the volcanoes are as large as any of the present-day ones.

These early Cascade volcanoes quickly eroded away after they became inactive. What remained was buried under younger volcanic rocks. However, in some of the deeper valleys, you still can find pieces of those old volcanoes.

One of the early Cascade volcanoes, the 23-million-year-old Fife's Peak volcano, can be seen a short drive up the Chinook Pass highway west of Yakima. The road takes you right through the heart of the volcano.

If you set your odometer to 0 (or just record the mileage) at the intersection of U.S. Highway 12 (the White Pass highway) and the beginning of the Chinook Pass highway (State Route 410), you can take a leisurely ride through the Fifes Peak volcano and use the mileage to find everything described in this column.

From the intersection, drive 14.1 miles and you are at the first place you can see the Fife's Peak volcano. Here, the younger Columbia River basalt laps up on the eroded flanks of the volcano.

This is the beginning of our trip through the volcano. It will last only seven miles (from 14.1 to 21 miles). If Mount Rainier were sliced in half, a similar trip through that slice would be about 10 to 12 miles. This comparison gives us an idea of how much of the base of the original volcano remains.

For about the first mile and one-half, you pass through the eroded southeast flank of the volcano. The rocks are brown to reddish brown and look broken. These are lava flows from the volcano that quickly cooled after they erupted. The solid parts broke as the lava continued to flow down the side of the volcano.

Two and one-half miles from where we entered the volcano (16.6 miles from the U.S. Highway 12/State Route 410 intersection), the rocks look different. There are shades of green and yellow throughout the rock, and the rock looks rotten.

This is an even older volcanic rock that underlies the Fife's Peak volcano. If you look closely, you'll see the Fife's Peak rock over the older volcanic rock higher up the hill.

The older green and yellow rock is lava from one of the earliest Cascade volcanoes. This older volcano, the Stevens Ridge volcano, grew to the size of Mount Rainier, and then became inactive. By the time the Fife's Peak volcano started erupting, the Stevens Ridge volcano was almost completely eroded away.

For the next three miles (16.6-19.2 miles), you will be driving through the Stevens Ridge volcano. Then you should notice the color of rock changing back to the brown to reddish-brown color of the Fife's Peak volcano.

At 19.2 miles, you're back in the Fife's Peak volcano but on the northwest flank; the center of the volcano is off to the south. If you look up to the cliffs, you'll see the rocks seem to have a steep inclination in the direction you're traveling.

Geologists have tried to reconstruct the Fife's Peak volcano by comparing the inclination of these lava flows to modern volcanoes. They estimate that the Fife's Peak volcano exceeded 10,000 feet, making it comparable in size to Mount Adams or Mount Rainier.

When you get to Cliffdell (19.9 miles), you can see some long, narrow rocks cutting vertically up through the lava flows. These are "dikes," the remains of the conduits that allowed the molten rock to pass upward through the volcano and erupt as lava flows. At mile marker 21, you are back in the Columbia River basalt again and have reached the other side of the volcano.

This short drive gives you an idea of the life of volcanoes in the Cascade Range. The modern volcanoes are built on the skeletons of earlier ones. If you could take a field trip through the Cascades in 10 million years, what the Fife's Peak volcano looks like now is probably what Mount Rainier will look like then.

Clear Creek Falls Scenic Viewpoint

About two miles from White Pass Village on the east side of the Cascade Range is the little known scenic viewpoint of Clear Creek Falls. The scenic view is hidden in the trees on the south side of the road just before it descends to Rimrock Lake and Yakima. Those who have taken the time to stop have been well rewarded by its spectacular view, in spite of the restrooms.

The view is well hidden by the trees. You have to follow one of the trails 10 yards east to the fence. From the fence, you can see Clear Creek Falls where it drops several hundred feet into the broad, steep-walled valley below.

The waterfall is a great example of how nearly 200 million years of geologic evolution produced a viewpoint so pleasing to the eye that you even forget about the cold and snow in winter. The geologic story of the viewpoint starts nearly 300 million years ago. This was just before the age of dinosaurs.

At that time, one of the tectonic plates began moving a piece of the Earth toward Washington, which was near the equator then. When this plate collided with the North American plate about 90 million years ago, it plastered a piece of land onto the North American continent.

The rock that makes up Rimrock Lake and Clear Lake Valley is one such piece of land. You can easily see this rock along U.S. Highway 12 above Rimrock Lake. The rock is easily recognizable because it usually has light green and grey colors. It also is weak and erodes to make the wide valley floor. Compare this to the narrow and steep Naches River Valley below the dam, which cuts through hard volcanic rock.

Above the old, weak rock of the valley floor is young, hard volcanic rock that forms the valley walls. This is called the Spiral Butte lava, and the volcano is Spiral Butte, just across the road from the scenic view. The Spiral Butte lava erupted a little over a million years ago and covered most of the nearby area.

After the volcanic eruptions ended, glaciers stepped in to write the last major chapter in the story of Clear Creek Falls. Rimrock Lake lies in a U-shaped valley, a valley with steep walls and a broad floor. Glaciers carve valleys like this. Clear Creek Falls is at the head of this U-shaped valley.

During the Ice Age, glaciers from White Pass moved east and sculpted the valley. After the glaciers melted away, the steep valley walls and wide valley floor were left behind. Clear Creek began flowing east from White Pass about 10,000 years ago, after the glacier retreated from the valley. This was the birth of the waterfall. As you can see, it takes a few hundred thousand millennia to make a viewpoint.

Satus Pass

Writing about areas of higher elevations in the winter time usually means writing about snow and cold. However, in the Tri-Cities, when the cloud inversion months set in, many of the higher elevations around us are above the fog. These are warmer places where you can see the sun.

The Horse Heaven Hills is one of the closest places to the Tri-Cities where you can occasionally escape the winter fog. On a good day, you can break through the fog on McBee Grade above Benton City, at the pass on State Route 221 above Prosser, and a little farther west along the Mabton-Bickleton road.

West of us, the Horse Heaven Hills gradually climb in elevation toward Satus Pass on U.S. Highway 97 and the Cascade Range. At 3,100 feet, Satus Pass is the highest pass across the Horse Heaven Hills. This extra height gives it the best chance of having clear skies when the rest of us are in the fog.

Like all the ridges that surround the Tri-Cities and lower Columbia Basin, the geology of the Horse Heaven Hills is a story of lava eruptions followed by buckling of the lava flows as they were squeezed from the north and south. The Satus Pass area shares this history but it got a little extra boost in elevation from another geologic event that is unique to that part of the Horse Heaven Hills.

The bedrock of Satus Pass is the same basalt lava that makes up most of the Columbia Basin landscape. After the lavas hardened into rock, Earth forces, which still operate today, compressed the region from the north and south. The rock responded by buckling into a series of east-west ridges. Rattlesnake Mountain, Badger Mountain, Red Mountain, and the Horse Heaven Hills are some of the more familiar names of these ridges.

Rattlesnake Mountain and the Horse Heaven Hills are higher than most ridges around here, but when you take a close look at them, the higher elevations typically are on the north side. The north side is highest because it's the point where the rocks change from a north facing slope to a south facing slope. If you drive from Prosser to Paterson on State Route 221, you go up a steep grade to the top of the hill and start a 35-mile gentle drop in elevation to the south.

Satus Pass is on the north side of the Horse Heaven Hills just like McBee Grade and the overlook above Prosser on State Route 221. However, on a clear day, the Satus

Pass portion of the Horse Heaven Hills clearly stands above the surrounding area. This portion of the Horse Heaven Hills is such a prominent landmark that it is called the Simcoe Mountains on maps.

To a geologist, the Simcoe Mountains are more than just a high place on the Horse Heaven Hills. Volcanoes dot these mountains. About 5 million years ago, magma began pushing up into the crust of the Earth below this part of the Horse Heaven Hills. The magma eventually broke through to the surface and began erupting lavas and forming cinder cones. The volcanism persisted until about a million years ago when the last volcano erupted lava that flowed down off the Columbia Hills and into the Columbia River near the Maryhill Museum.

Today, the only remaining products of the earlier volcanism are the cinder cones that are mined for the red lava rock we see in many lawns around the Columbia Basin and that pleasant, sunny warm pass at the top of the fog.

Maryhill Museum

Maryhill Museum is one of the hidden gems of the Columbia Basin. It overlooks the Columbia River just east of U.S. Highway 97 and offers a variety of geologic features in addition to its history and art collections and the scenic view.

The museum sits on a bench 800 feet above the north side of the Columbia River along State Route 14. North of the museum, the Columbia Hills rise to an elevation of nearly 2,800 feet and overlook the Columbia River Valley below. The perfect setting for the museum owes its existence to the geologic history of the area. From the museum, you can see examples of three major processes that helped shape the area over the last 15 million years: volcanism, faulting, and cataclysmic floods.

The Columbia River began occupying this river channel less than 7 million years ago when it began using Wallula Gap and abandoned a previous course that passed by Goldendale. The valley below Maryhill received its most recent sculpturing by the Ice-Age floods of the Pleistocene. After these floods carved the channeled scablands of eastern Washington and drained through Wallula Gap, the water was channeled into the narrow gorge below Maryhill. No one knows how much the river channel was widened by the floods, but they certainly filled the valley, and you would have gotten wet if you had been standing at the museum's site.

Maryhill Museum sits on a bench made of nearly flat lying lava flows of Columbia River basalt. From the river up to the bench, U.S. Highway 97 twists through lava flows that are about 15 million years old. The museum itself sits on the 14.5-million-year-old Priest Rapids lava flow named after Priest Rapids Dam.

The lava flows below the museum show some of the best examples of columnar jointing that can be found around the Columbia Basin. When you drive up U.S. Highway 97 from the Columbia River to State Route 14, you pass stately columns several feet thick that extend for miles along the river.

These same lava flows sit 1,000 feet higher on top of the Columbia Hills, but they don't look the same as they do near the river. If you were to walk along State Route 14 west or east of the museum, you would see that the rock is broken and shattered, which is very different from the solid rock the museum sits on. This broken rock extends up the Columbia Hills and is exposed at many places along U.S. Highway 97.

The broken rock is a clue that tells geologists that tectonic forces were active here in the past. This zone of broken rock marks the Columbia Hills fault zone. The rock north of State Route 14 has been forced up over the rock that forms the bench under the museum and road. This fault zone can be traced from near Hood River, Oregon, along the north side of the Columbia River to as far east as Plymouth, Washington, and Interstate 82. As the Columbia Hills grew, rock moved along this fault, and the Columbia River was forced into its channel below.

If you go west along the bench and State Route 14 for a couple of miles toward the beginning of the Columbia Gorge Scenic Area, you will see Miller Island in the Columbia River. On the island, there is some unusual gray rock that is not like Columbia River basalt. This is a young lava flow, about 900 thousand years old, that erupted from a volcano that sits directly on top of the Columbia Hills. The lava flowed down the hillside, across the bench, and down into the river and the island.

The 900-thousand-year-old lava is part of the Simcoe lava field that produced lots of cinder cones around the Goldendale area. A lot of the red rock for gardens in the Tri-City area comes from these cinder cones. The young lava covers the Columbia Hills fault zone and doesn't appear to be faulted, which suggest this fault zone hasn't moved since that lava flow covered it.

Maryhill Museum is about a 2-hour drive from the Tri-Cities. To get a real feel for this area, drive Interstate 82 south to the Washington-Oregon border, then follow State Route 14 along the Columbia River to U.S. Highway 97. This route follows the south flank of the Columbia Hills and the same fault zone you see at the Maryhill Museum. As you drive this road, watch how the Columbia River follows the Columbia Hills and imagine how it must have looked with the cataclysmic floodwaters completely filling the channel and rushing toward what is now the Maryhill Museum.

Corps of Discovery

On October 28, 1805, Lewis and Clark and their Corps of Discovery woke to "a cool windey morning" and began the last part of their journey to the Pacific Ocean. They had spent that night along the shore of the Columbia River near The Dalles, Oregon.

Within a few miles, Clark noted in his journal that "we proceeded on river enclosed on each Side in high Cliffs of about 90 feet of loose dark coloured rocks." They were now entering the Columbia Gorge at a place we now call Rowena Gap.

Nearly 170 years later, I first saw this place while on a geology field trip lead by Dr. Aaron Waters, one of the pioneering geologists of the Pacific Northwest. We stopped there to study the geology that Lewis and Clark had seen because that area had become important to geologists studying Pacific Northwest history.

A while back, almost 30 years after I first saw Rowena Gap, I took a colleague from France to that part of the "Louisiana Purchase" to see the geology. He is a planetary geologist and has spent much of his career studying the geology of other planets in our solar system. He wasn't so interested in what his country had sold us long ago, but he wanted to see the "loose dark coloured rock" that Lewis and Clark had described in their journals.

Rowena Gap is an attraction to geologists not only because it is the gateway to the Columbia Gorge but because it is one of the best places to see into the heart of one of the ridges that dominates the Columbia Basin: the Columbia Hills.

The Columbia Hills are a landmark along the Columbia River and figured prominently in Lewis and Clark's journals. The ridge stretches from Interstate 82 where it crosses the Columbia River at Umatilla to the Cascade Range. For most of that distance, it forms the north shore of the river.

However, between Lyle, Washington, and The Dalles, Oregon, the Columbia River makes an abrupt turn to the northwest and cuts right through the Columbia Hills. Lewis and Clark noted this large departure from the overall southwest flow of the river.

Rowena Gap provides geologists with a spectacular view of the Columbia Hills. From the east, the layers of basalt lava rise rapidly westward. Erosion has selectively removed the lava in such a way that each lava flow forms large, distinct, and tilted layers on the walls of the gorge. Then, abruptly, the layers steepen to vertical and drop straight into the river.

This bending of the layers forms a geologic feature called an anticline. Faults, where the layers break and separate, bound each side. The fault on the east side isn't easy to see, but the one on the west side is spectacular! This is where the layers of lava dive vertically to the river. You can see the vertical layers of basalt contact with horizontal layers of basalt. You'll never see a better example of a fault anywhere in the world.

Why, you might ask, did this geologist who studies planets travel from Europe to Rowena Gap? The answer is in my colleague's own quest for discovery. The lavas and the ridges, like the Columbia Hills, that form the Columbia Basin are uncommon on Earth but are actually quite common on planets like Mars and Venus. The lava flows and ridges we live with every day are some of the most fundamental features of the inner planets of the solar system. Planetary scientists like my French colleague have discovered that our little part of the Louisiana Purchase is an important key to understanding how our solar system evolved.

Beacon Rock State Park

On the north side of the Columbia River just west of Bonneville Dam, Beacon Rock stands like a lighthouse built to guide travelers through the Columbia River Gorge. This is probably one of the most easily recognized and photographed features in the gorge and is the centerpiece for the beautiful state park that surrounds it.

On October 31, 1805, Lewis and Clark, on their Corps of Discovery, noted this landmark in their journals. They called it "Beaten Rock," but then later corrected it to Beacon Rock. It was called Inshoack Castle by the John Jacob Astor expedition in 1811 and also was known as Castle Rock until Henry Bittle purchased it in 1915 to save it from becoming a quarry. A year later, in 1916, its original name was restored.

Beacon Rock became a state park in 1935 only after ownership was transferred from the state of Oregon to Washington. Bittle's heirs deeded the landmark to Oregon in 1932 to protect it from Washington's governor. Although many people in Washington wanted it to become a state park, the governor at that time, Ronald Hartley, didn't believe in state parks. After a new administration took over in Olympia, the state of Oregon gave Beacon Rock to the people of Washington for a park.

In addition to its historic importance, Beacon Rock also is a centerpiece for the geology of the Columbia Gorge. It is an 840-foot-high remnant of an eroded volcano.

The Cascade Range is made up mainly of volcanic rocks. For mountain ranges in the United States, it is one of the younger ones, starting to grow only about 38 million years ago. The modern Cascades are typified by the high Cascade volcanoes such as Mount Rainier and Mount Adams, which are less than 1 million years old.

Beacon Rock is the remains of a volcano that erupted about 1.6 million years ago. This was long before mankind arrived, but only yesterday to the Cascades.

Volcanoes erupt lava through a central conduit called the neck. As lavas erupt through the life of the volcano, a cone is built up around the neck. The cone of the volcano is usually more easily eroded than the neck. Lava in the neck is protected by the cone, and when it cools, can become a hard, massive rock. Beacon Rock is the neck of the volcano.

The volcano at Beacon Rock is just one of several volcanoes that fed lava flows of the same age throughout western Oregon. Geologists call these the Boring lavas, not because they are a boring rock to look at, at least to a volcanologist, but because they were first described at the town of Boring south of Gresham, Oregon.

The Beacon Rock volcano is different from most other volcanoes because it had the misfortune to erupt right in the path of the Columbia River. During eruptions, lava from the volcano would periodically dam the river for a short time. The volcano, however, was no match for the river. The lava and the volcano were just minor inconveniences that were easily eroded away. Now, all that remains of the volcano is the neck. The Columbia River was able to remove all but the hard rock of the volcano's neck. Given enough time, the river will remove that as well.

Next time you are in the Columbia Gorge, plan a visit to Beacon Rock. Henry Bittle built steps to the top of the rock. Not only do they lead to a great view of the gorge, but you can also walk up the center of a volcano.

Bridge of the Gods

A Native American story about the Bridge of the Gods in the Columbia Gorge says the Great Spirit had two sons: Wy-east (Mount Hood), who lived in the Cascade Range south of the Columbia River, and Pah-toe (Mount Adams), who lived north of the river. All was well until a beautiful maiden moved into the valley between the two brothers. Both brothers fell in love with the maiden and began to fight over her. Legend has it that they stomped the ground causing it to shake violently; then they threw rocks and fire at each other. Their fighting brought destruction to the nearby land and all who lived there.

The fighting also triggered a landslide that dammed the Columbia River. After a while, the Columbia River was able to tunnel through the dam forming a natural bridge: the Bridge of the Gods.

When the Great Spirit learned of the fighting, he promptly stopped it. He sent the maiden away, and proclaimed that the bridge would remain as a symbol of peace between the brothers. However, if the peace ended, the bridge would be destroyed.

One of the brothers (Mount Hood) was able to find the maiden again and began to romance her. When the other brother found out, he began tossing fire and rock at the other. The fighting shook the ground so hard that the bridge collapsed into the river.

This legend always has intrigued geologists because it's clear that it grew out of stories by those who saw the Bridge of the Gods area form. Cascade Locks at the Bridge of the Gods was notorious before the construction of Bonneville Dam for the "Narrows" and "Cascades of the Columbia" first described by the Lewis and Clark Corps of Discovery in 1805. The locks were constructed to help river traffic bypass this obstacle.

Geologists studying the area recognized that the Narrows and Cascades of the Columbia were caused by a landslide that blocked the Columbia River. The remnant of the landslide covers nearly 50 square miles on the north side of the river where the slide originated. Age dating of trees caught in the landslide indicates that most landsliding occurred about 1260 A.D.

The geologic story of the Bridge of the Gods begins in the Pleistocene Epoch when the ice-dammed lake near Missoula, Montana, released great torrents of water across Washington. This water, the Missoula floods, drained through the Columbia Gorge and out to the Pacific Ocean.

When the floodwaters raced through the Columbia Gorge, the water cut away soft rock on the north side of the river leaving steep walls.

The sons of the Great Spirit in the Native American legend are the Cascade volcanoes: Mount Adams and Mount Hood. Both volcanoes are young and active, and their eruptions have been witnessed many times by native peoples. The fire and rocks of legend describe the volcanoes erupting, but there is no geologic evidence to show that they erupted when the landslide occurred. Mount Hood erupted around 995 A.D. and 1760 A.D., but Mount Adams has not had a major eruption in 3,500 years.

The brothers stomping the ground during their fight were the culprits who caused the landslide. Most geologists think the landslide was triggered by a major earthquake caused by the two major tectonic plates colliding along western Washington and Oregon.

Colliding tectonic plates cause large earthquakes on the average of every 200 to 400 years. The steep walls of the Columbia Gorge formed by the Missoula floods, weak rock, and the high rainfall of the area made the perfect setting for landsliding. The ground shaking caused by a big earthquake is more than enough to let slide already weakened rock.

The landslide dammed the Columbia River, possibly for several years, and caused a lake to form. There isn't any geologic basis for a bridge other than the dam, however. Considering the nature of the material forming the landslide, the rock wouldn't have been competent enough to form a bridge if the Columbia River had undercut it.

The landslide dam was as much as three times higher than Bonneville Dam but, unlike Bonneville Dam, when the lake reached the top, it overflowed and rapidly cut a new channel, the Narrows, and the Cascades of the Columbia.

Many legends of the Pacific Northwest, like the Bridge of the Gods, have strong foundations in the natural history of the Pacific Northwest. These legends make us realize that the first inhabitants of the area also were the first geologists, and much of what we have learned about the area, they already knew.

Lake Coeur d'Alene

When those of us living in the Columbia Basin think of a large lake, we usually think of those that formed behind the dams along the Columbia and Snake rivers. Many small lakes are scattered throughout the basin, but you'll be hard pressed to name many large lakes that aren't man made. Even many of the large lakes in the Cascades such as Rimrock or Bumping Lake sit behind a man-made dam. This isn't surprising to a geologist, because a natural lake is a short-lived feature. Natural lakes usually are doomed to a few tens of thousands of years at most.

Some of the largest lakes in Idaho, however, were formed by natural processes. The second largest lake in Idaho, Lake Coeur d'Alene, is a good example. Lake Coeur d'Alene lies in northern Idaho just east of Spokane, Washington. It stretches for nearly 50 miles from Interstate 90 south to St. Maries, Idaho. Along most of its length, it is barely two miles wide. The lake is fed by both the St. Joe River and the Coeur d'Alene River from the east and the St. Maries River from the south. The city of Coeur d'Alene sits at the northern end of the lake, which is the headwaters of the Spokane River.

If you examine the land surrounding the lake, you probably would guess that someone simply dammed up the Spokane River. The lake lies in a beautiful valley that extends south into the Bitterroot Mountains, and it is dammed right where Interstate 90 follows the Coeur d'Alene Valley across Idaho, a natural place to put a dam and make a lake!

This same thought must have crossed Mother Nature's mind because it was she who built the dam. Lake Coeur d'Alene is a consequence of the Pleistocene Epoch (the Ice Age). We have to go back nearly two million years to reconstruct the history of Lake Coeur d'Alene.

Long before the Pleistocene began, the St. Joe River flowed north to the site of the present city of Coeur d'Alene. However, rather than join the Spokane River and flow west, it continued north. The St. Joe River flowed through the Rathdrum Prairie, the large valley north of Coeur d'Alene and then turned west and joined the Pend Oreille River and eventually the Columbia River.

During the Pleistocene, many continental glaciers advanced south from Canada. Between 130,000 and 70,000 years ago, a lobe of an ice sheet moved down the Purcell Valley from Canada and south across the Rathdrum Prairie. The Purcell Valley is the large north-south valley that U.S. Highway 95 follows north into Canada.

This glacier pushed south as far as the present city of Coeur d'Alene and then stopped. As it came across the Rathdrum Prairie, it blocked the St. Joe River, forming Glacial Lake Coeur d'Alene.

If the ice were the only thing damming the lake, the St. Joe River would have just reoccupied its valley after the ice melted. However, glaciers are also Mother Nature's bulldozers. As the glacier pushed down from Canada, it also carved out valleys and moved lots of rock.

At the end of a glacier, there is usually a large pile of rock and dirt that is pushed along by the glacier. Geologists call the rock and dirt "till." The pile of till at the end of a glacier is called a terminal moraine. It's the terminal moraine that continued to block the path of the St. Joe River long after the glacier melted and created present-day Lake Coeur d'Alene.

The Spokane River also formed at about the same time. Lake Coeur d'Alene was always kept full from the rivers flowing into it. After it filled, it overflowed its natural dam, forming the Spokane River. The terminal moraine and surrounding hills forced the Spokane River to cut its present valley to the west to join the Columbia River.

Steptoe Butte and Old North America

Steptoe Butte is probably the best known landmark in the Palouse country of eastern Washington. It is a conical-shaped mountain that rises to an elevation of 3,612 feet. This is about 1,000 feet above the surrounding countryside. You can see Steptoe Butte from just about any vantage point in the Palouse and even as far south as the Blue Mountains.

Native Americans called it the power mountain, but its current name was derived from Colonel Edward Steptoe. The butte was used as a lookout by the Army, and the battle of Rosalia was fought nearby between the Army and the Native Americans. Now, just about all similar looking features in eastern Washington are called "steptoes."

To geologists, Steptoe Butte is a window into a little known part of the geologic history of the Pacific Northwest. The main rock that forms Steptoe Butte is quartzite. Quartzite is a metamorphic rock that originally was composed of quartz sandstone. The sandstone was subjected to heat and pressure (metamorphosed) to the extent that the quartz sand grains fused together to make a very hard rock.

The quartzite of Steptoe Butte also contains the mineral mica. Mica is best known as the mineral that peels apart in sheets. Old wood-burning stoves had isinglass windows made of mica, as did some early cars. The mica formed from clay when the sandstone was metamorphosed.

The rock of Steptoe Butte is more than 800 million years old, which makes it one of the oldest rocks in eastern Washington. Geologists call these rocks the Belt Supergroup, and more specifically, the Ravalli Group. These geologic names are derived from geographic features of Montana, specifically, the Belt Mountains and Ravalli County where they were first recognized. Glacier National Park is one of the most famous places to see rocks like these.

Besides being old, these rocks tell us part of the story of how the Pacific Northwest formed. Before 800 million years ago, the rocks of Steptoe Butte were being deposited as sediment in a large basin that extended east into central Montana and north into Canada, as well as south and west. The basin was in the large supercontinent called Rodinia. Think of Rodinia as a large continent formed from all the continents on Earth.

About 800 million years ago, the supercontinent of Rodinia broke apart. The boundary of one of those continents breaks north-northwest and lies under the

Tri-Cities. As pieces of Rodinia moved apart, the Pacific Ocean was born. The piece of Rodinia we were joined to is now in Australia and Asia. Just think, if you had been here 800 million years ago, you would have lived on the shore of a new ocean.

Needless to say, quite a lot has happened to Washington since that time. All the land west of us had to have been added on over the past 200 million years. In addition, younger sediments and volcanic rocks covered all the older rocks.

Steptoe Butte remains one of the last remnants of a vast mountain range that once existed in eastern Washington. These mountains were slowly eroded away and then covered by the younger rocks. If it weren't for isolated rocks like Steptoe Butte that poke above the younger rock, the early history of Earth would be an even bigger mystery.

Petrified Wood at Palouse Falls

Palouse Falls, located at Palouse Falls State Park northeast of the Tri-Cities, is a spectacular 185-foot waterfall that overlooks a 400-foot-deep canyon of the Palouse River. The most famous part of the park's geologic story is that of the Ice-Age floods and how those floodwaters rerouted the Palouse River and carved the canyon and waterfall. However, there is another story hidden in the rocks that is less well known.

The Palouse River originates in the mountains of Idaho and flows through Pullman and Colfax then west across the Palouse country as far as the town of Hooper on State Route 26. Just west of Hooper, the river makes an abrupt bend to the south and joins the Snake River at Lyons Ferry. Palouse Falls State Park is five miles upstream from the confluence of the Palouse and Snake rivers.

The river didn't always follow this course, however. Before the great Ice-Age floods, the river followed a channel down Washtucna coulee past the present towns of Washtucna, Connell, and Mesa, and then down Esquatzel coulee past Eltopia. It probably joined the Columbia River not far from the location of the Pasco Farmer's Market.

The Ice-Age floodwaters cut the canyon to Palouse Falls in several days. Since then, the waterfall has been in nearly the same place. Another fortunate consequence of the tremendous erosive power of the floodwaters was the exposure of rock along the sides of the canyon. This rock consists of layers of Columbia River basalt lava flows that are about 14 to 15 million years old. The individual lavas stand out as steep slopes and the contacts between the lavas are small benches.

The waterfall occurs just below one bench and exposes an important contact between two major groups of lava flows along that bench. This contact marks the end of the greatest period of lava production when over 40,000 cubic miles of lava were erupted in just one million years. After that, there was a hiatus in volcanic activity that lasted for as much as 250,000 years. Sediment was deposited during this hiatus before volcanic activity resumed. This younger volcanic activity occurred at a much slower pace, however, accumulating a mere 2,500 cubic miles of lava in one million years.

The sediment layer at Palouse Falls also represents the same time period that petrified wood occurs at the Ginkgo Petrified Forest at Vantage. The sediment at Vantage is from the ancestral Columbia River, but the sediment at Palouse Falls is from some river that flowed from the east or northeast. Perhaps it was the ancestral Palouse River 15 million years ago.

The nice surprise at Palouse Falls is the presence of petrified wood and charcoal from plants that were growing in the sediment when the lava buried it. The sediment layer also played an important role in the more recent history of the Palouse River. If you walk north from the falls along the riverbank, you are following this horizon. The sediment layer forms a weakness in the layers of rock and as the Ice-Age floodwaters were moving down the valley, the rock above the sediment was more easily removed.

To get to Palouse Falls, drive to Connell along U.S. Highway 395. When you get to Connell, turn east on State Route 260 and follow it past Kahlotus. This is the old Palouse River Valley, but the present valley was deepened and enlarged by the floodwaters. The turn off for the park (State Route 261) is 8 miles beyond Kahlotus and is marked by signs for Palouse Falls and Lyons Ferry. State Route 261 will take you out of Washtucna coulee, over a hill and into the first dry channel. Note the dry falls (cataract) south of the road. The road then goes over a second hill and into another wide valley. When you start down the gravel road into the park, you will see the inner gorge the Palouse River follows.

Swallows Rock

The twin cities of Lewiston, Idaho, and Clarkston, Washington, at the confluence of the Snake and Clearwater rivers are the gateway to Hells Canyon. The water impounded behind Lower Granite Dam makes the twin cities a port, but once you go up river of the confluence, very quickly you are in jet boat country full of fast-flowing water and rapids.

The two cities derive their names from Lewis and Clark's Corps of Discovery. The two early explorers passed by the future cities on their way down the Clearwater to the Snake and on the overland route on their way back east. The Lewis and Clark journals are full of descriptions of the area, but the one thing they missed is the landmark that stands out to all the residents: Swallows Rock.

Swallows Rock is just up the Snake River from Clarkston on the Washington side of the river. It looks like a Pacific Northwest version of the guardian rock of the Mediterranean Ocean: the Rock of Gibraltar. A local resident once told me that the Prudential Insurance Company, which uses the Rock of Gibraltar as its symbol, really uses Swallows Rock instead because it was cheaper for the company to come to Lewiston. I don't think anyone really believes that story, but the point is well taken. Just like the Rock of Gibraltar, Swallows Rock stands as a guardian.

The geologic story behind Swallows Rock is one of basalt, rivers, and erosion. Like most of the Columbia Basin, the rock that makes up Swallows Rock is Columbia River basalt. There are two lava flows that combine to form the landmark: the 12-million-year-old Pomona lava and the 10.5-million-year-old Elephant Mountain lava.

Before the Pomona and Elephant Mountain lavas erupted, the rivers had cut down several hundred feet through layers of earlier basalt lavas. Geologists have traced this formidable river canyon as far west as Goldendale. During this time, the Lewiston-Clarkston area had begun to develop into the present Lewiston Basin where both cities now are located.

Both lava flows erupted from fissures in the Lewiston-Clarkston vicinity. The Pomona erupted farther up the Clearwater River near Orofino, and the Elephant Mountain erupted from farther up the Snake River. Both lavas spread throughout the Lewiston Basin, but the only way out was the channel carved by the combined rivers. Swallows Rock sits in the heart of that old river canyon.

The first lava to use that canyon was the Pomona lava. It flowed down the ancestral Clearwater River, spread out in the Lewiston Basin, and then flowed west from the basin. This lava flow made it all the way to the Pacific Ocean. In the Tri-Cities area, we can see the lava flow along U.S. Highway 395 near Mesa, on the ridges of the Tri-Cities, and along the Yakima River between Horn Rapids and Benton City.

The Pomona lava probably filled the canyon to the brim, but the ancestral rivers quickly reestablished themselves in that old canyon. By the time the Elephant Mountain lava erupted 1.5 million years later, the lava easily could flow down the same channel used by the Pomona lava. The remnants of the earlier Pomona lava formed the bed of the river canyon.

The Elephant Mountain lava almost made it as far as the Pomona lava. We can see it capping most ridges around here and along the Yakima River between Benton City and Sunnyside. The Elephant Mountain lava proved too much for the rivers, however. It filled the old canyon to overflowing. Rather than cut back into the old canyon, the rivers abandoned the channel for the current one used by the Snake River. For the Snake to join the Clearwater River again at the present confluence, it had to clear out all the lava blocking its path.

The steep river side to Swallows Rock is all that remains of the two lavas. Like the guardian of the canyon, it stands as a monument to the impressive size of the basalt lavas and the even more impressive erosional power of rivers.

Fields Spring State Park

Field Springs State Park offers one of the most spectacular views in southeastern Washington. The park is located about 25 miles south of Clarkston. It overlooks the Grande Ronde River canyon, a 3,000-foot-deep canyon cut into the Columbia River basalt. The main feature of the park is Puffer Butte, the remnants of an old volcano for one of the lava flows.

Puffer Butte is on the south rim of Grande Ronde River canyon where Washington State Route 129 climbs down Rattlesnake Grade to the river and the famous Bogan's Oasis. It's about a 3.5-hour drive from the Tri-Cities, so you might ask: Why go there? Because of the geology, of course!

The park is open year-round and offers camping, hiking, and cross-country skiing in addition to the fantastic view of the canyons. The park also is the birthplace of a lava flow that is an important part of our own region. Our local geology can trace some of its roots back there.

The park covers about one-half square mile, with the campground nestled in the trees at Fields Spring. Trails cover the park and the rim of the canyon and even drop down to the river far below, giving you the chance to walk through rock that is well over a half-mile below us in Tri-Cities.

The geologic history of Fields Spring Park began about 15 million years ago when lavas began erupting through cracks or fissures in the ground. These are still preserved as lava-filled fissures called dikes. By 13 million years ago, a vast plain of lava had built up. Then a fissure opened up at Fields Spring, and lava started pouring out. By the end of the eruption, a small volcano had built up over the fissure. This is Puffer Butte, and the lava flow is called the Umatilla flow.

The Umatilla lava spread north and south but was not able to go too far east because of the foothills of the Bitterroot Mountains. The main route west for the lava turned out to be south and west across what had to have been a divide across the Blue Mountains. This is near where the Oregon Trail crossed the Blue Mountains. This divide was the channel of the ancestral Clearwater and Salmon rivers that flowed from the Bitterroot Mountains to the Pacific Ocean.

How do we know the lava followed this southerly course? The answer was discovered about 20 years ago near Tollgate, Oregon, and Spout Springs Ski Area in the Blue

Mountains. The highest part of the present ski area is a remnant of the lava from Puffer Butte overlying river gravels!

The lava flooded west through this river channel into the Columbia Basin and the Tri-Cities, reaching as far north as the Hanford Reach of the Columbia River and almost as far west as Goldendale.

If you drive back to the Tri-Cities via Enterprise or Elgin, Oregon, to Tollgate and Milton-Freewater, you'll be retracing the path followed by the Umatilla lava. Your driving time will be much quicker, however; the Umatilla basalt took a week or more to go the same distance.

You can reach Fields Spring State Park by going east on U.S. Highway 12 to Clarkston and then turning south to Asotin, State Route 149. For the more geologically inclined, a geologic map of the area (GM-40) is available from the Washington Division of Geology and Earth Resources, P.O. Box 47007, Olympia, Washington 98505-7007.

La Grande Basin

About 60 miles southeast of the Tri-Cities is a beautiful little valley nestled between the Blue Mountains and the spectacular Wallowa Mountains. This is the La Grande Valley, the valley of the Grande Ronde River.

The La Grande Valley lies just about where the younger rocks of the Columbia Basin to the north lie against older rocks of Oregon that became part of North America when the dinosaurs walked the Earth. It is the northwestern-most of a series of valleys that Interstate 84 passes through beginning at Farewell Bend on the Snake River. These valleys are in the heart of a major northwest-oriented fault zone, and it is these faults that give the valley its shape.

The valley formed when two different fault directions started moving. One was the major northwest trend, and the other was a north-northeast fault that ripped the valley in half. As faults ripped the valley open, volcanic rocks filled in the valley floor.

The southwest side of the valley that Interstate 84 follows south through La Grande to Union is the main fault where rocks slid by each other on the south side. The northeast side of the valley, just north of the town of Imbler, is the other major northwest fault. The walls of the valley that trend north from La Grande on the west and Union on the southeast once were joined.

The La Grande Valley received attention a few years ago because the U.S. Army Corps of Engineers found evidence of a very young fault in the valley. This fault is thought to be a sign that the area is active and may be capable of producing large earthquakes. This is not a new idea. The largest earthquake we have had in the Columbia Basin in recent times was the 1936 Milton-Freewater earthquake, which was almost a magnitude 6 and did a lot of damage to Walla Walla and the surrounding area.

The history of the La Grande Valley has become more clear in the last several years. It is now known that the valley began forming when the Columbia River basalt was erupting and laying down huge lava flows throughout the Columbia Basin. As the valley formed, the lava filled it in quickly, masking its formation. It was only after the huge lava flows of the Columbia River basalt ceased erupting that the valley development became obvious. The youngest volcanic rocks, the Powder River volcanics, were the last lavas to spread in the valley. These ceased erupting about 13 million years ago. It was thought that the valley had not changed much since those eruptions.

Recently, however, research has shown that the valley didn't stop forming then but continued to grow. We now know that the La Grande Valley was sinking and the hills rising at a combined rate of about 190 meters per million years. We think the valley is still growing at that rate today.

Besides that very young fault, there is other evidence for young fault movement in the valley. Valley residents know this evidence only too well. All around the valley walls along the boundary faults are many hot springs and water wells with warm-to-hot water! When faults move, they generate a lot of frictional heat. The presence of this hot water is one of the surest signs of an actively growing valley.

You easily can see the faults that formed the valley. As you drive southeast from the Tri-Cities to La Grande, look at the rock alongside the road as you near the La Grande Valley. You see Columbia River basalt broken by many faults. These are the northwest faults. When you get in the valley, look north to Mount Margaret and east to the Wallowas. Notice how sharp the valley walls are. These walls once were joined.

Fossils of Fossil, Oregon

When the town of Fossil, Oregon, was excavating for its new Wheeler High School in 1949, construction uncovered what was to become one of the most famous fossil localities in the Pacific Northwest. Since that initial excavation, the north end of the athletic field has been a major fossil collecting site.

Fossil, Oregon, is located in north-central Oregon south of Interstate 84 on State Route 19 where the road begins to descend into the John Day River Valley. The hills around the town are capped by thin layers of Columbia River basalt, but the rock below contains rich fossil beds. The town of Fossil isn't named for this locality, but for a fossil locality in Hoover Creek near the original post office.

The rock at Fossil is called the John Day Formation and is part of the geologic ages called Oligocene and Miocene. The age of the rocks ranges between 37 and 19 million years old and predates the Columbia River basalt by several million years.

Over 35 different species of fossil plants have been found at the Wheeler High School collecting locality. Many more species are thought to occur there, however. In spite of the long history of collecting there, little has been written about the fossils, and most collecting has been done by amateur paleontologists. Most species found at Fossil are no longer native to the Pacific Northwest. Although some plants are now extinct, most can now be found in Asia and eastern North America.

The Wheeler High School fossil locality is thought to be Oligocene in age (37 to 25 million years old). The fossils are found in volcanic ash deposited in lakes that are probably closer to 37 million years old. The Wheeler High School fossil beds are a little older than the famous vertebrate fossil beds of John Day Fossil Beds National Park to the south, which are thought to be between 25 and 19 million years old.

Most fossils are deciduous trees such as maple, alder, and beech, but conifers, including dawn redwood (*Metasequoia*) and pine easily can be found. Dawn redwood was thought to be extinct, but plants were discovered in China in the 1930s. The fossil plants are preserved as impressions and carbon films in the rock. The types of fossils range from leaves to fruits, cones, seeds, and occasionally a flower.

The best way to find the fossils is to either look at the surfaces of the slabs of rock or take a chisel and hammer and break the rock along its natural partings. There are many vertical fractures in the rock, so you have to be careful when you find some of the larger leaves. Also, the fossils are delicate, and before you take them home, be sure to wrap them in protective cloth or you'll find yourself with a pile of rubble when you get there.

These and other fossils from the John Day Formation have given geologists insight into what northern Oregon was like millions of years ago. These rocks are from the birth of the Cascade Range 37 million years ago. Volcanoes in the Cascades erupted and their ash settled into basins in the ancestral Blue Mountains. The ash covered plants and animals and preserved them for collectors today.

Fossil Fish of the Grande Ronde Basin

A little over 100 miles southeast of the Tri-Cities, the Grande Ronde Valley nestles between the Blue Mountains and the spectacular Wallowa Mountains. The valley is more than one-half mile above sea level, and the surrounding mountains rise to as much as 10,000 feet.

The Grande Ronde Valley is the northwestern-most valley in a series of valleys that Interstate 84 passes through beginning at Farewell Bend on the Snake River. These valleys are in the heart of a major northwest-oriented fault zone, and these faults give the valleys their shape and define their history.

The Grande Ronde Valley formed when two different sets of faults started moving. One fault set is oriented roughly northwest and the other is oriented northeast. The valley grew by a combination of uplift of the surrounding ridges and sinking of the valley floor.

The Grande Ronde Valley started forming over 15 million years ago when the Columbia River basalt was erupting and laying down huge lava flows throughout the Columbia Basin. As the valley started to form, the lava filled it in quickly, covering the valley floor and the surrounding ridges. It was only after the huge lava flows of the Columbia River basalt ceased erupting that the shape of the valley became obvious.

The youngest volcanic rocks to erupt in the Grande Ronde Valley were the Powder River volcanics. By the time they were erupting, the valley walls were high enough that some of the lava was confined only to the valley floor. The last of these volcanoes ceased erupting about 9 million years ago. After the last eruption, however, the valley continued to grow. Then, only sediments filled in the valley floor. Water wells show that, in some places, almost 2,000 feet of sediment fills the valley floor.

During spring 2000, a water well being drilled into these sediments brought up more than just water. The well driller found fossil fish in the drill cuttings. The fossils eventually made it to the University of Michigan's Museum of Paleontology where they were identified.

Four species of fish were identified: pikeminnow, bullhead catfish, western sunfish, and mountain whitefish. A volcanic ash layer at about the same depth in the well allowed geologists to estimate the age of the fish to be 3.8 million years old.

The fossil fish are very similar to those collected from fossil localities on the White Bluffs on the Hanford Reach National Monument and on Taunton Bench just south of Othello. They are also similar to fish of the same age from the Snake River Plain near Boise, Idaho.

The fossil fish in the Grande Ronde Valley led paleontologists to conclude that approximately 4 million years ago the Grande Ronde Valley was part of a large system of lakes and rivers that connected the Columbia Basin to the Snake River Plain in Idaho.

If you look at a map of the Pacific Northwest, the only possible connection is the Snake River where it cuts through Hells Canyon. However, we know from detailed geologic studies that the Snake River didn't cut through Hells Canyon until 2 million years ago. A great divide existed just south of the Salmon River, and it took until 2 million years ago for the Salmon River to erode that high area. However, the location of the Snake River before it flowed through Hells Canyon has always been a mystery.

The Grande Ronde fossil fish present another geologic mystery for the geologists of the Pacific Northwest: Where was the 4-million-year-old connection between the Columbia Basin and the Grande Ronde Valley?

Geologists have long speculated that before the Snake River cut through Hells Canyon 2 million years ago, it flowed across the Blue Mountains from Farewell Bend, Oregon, to near Arlington, Oregon, west of Boardman. They speculated that the old channel might have paralleled the route followed by the Oregon Trail and now Interstate 84. Perhaps these fossil fish in the Grande Ronde Valley are clues to the mystery of the Snake River.

Mount Multnomah and the Three Sisters

The Three Sisters volcanoes in the central Oregon Cascade Range are often in the news. One spring there were about 100 small earthquakes that occurred three miles west of South Sister. The U.S. Geological Survey attributed the earthquakes to a "modest amount of magma" accumulating at a depth of about four miles. The magma also has caused about 10 inches of uplift there since 1997. Geologists are now trying to determine what this really means to the area. Are we seeing the beginning phase of a future eruption or is it simply just a normal event in the life of a young volcano?

The volcanoes that make up the Three Sisters are South Sister, Middle Sister, and North Sister, but they are just three of many more volcanoes that occur close together. Broken Top, Belknap Crater, Black Butte, Mount Washington, and Mount Bachelor are all young volcanoes nearby. The earliest known volcano that grew in the area was named Mount Multnomah by Dr. Edwin Hodge of the University of Oregon in 1925. He argued that this was one of the largest volcanoes in the Cascades, rising over one mile above the tops of the Three Sisters, but it blew apart creating one of the largest craters in the world. The Three Sisters sit on its remains.

The volcanoes of the Three Sisters area, along with Newberry volcano and Mount Jefferson, dominate the landscape of Central Oregon. These volcanic centers are pleasant companions to watch while driving from Madras to Bend, Oregon, along U.S. Highway 97. The Three Sisters are especially important volcanoes because they lie in that part of the Cascades directly west of the rapidly growing areas of Redmond and Bend. But even more importantly, these volcanoes erupted repeatedly for hundreds of thousands of years and into recent time.

North Sister is the oldest of the Three Sisters. She sits on a 200,000-year-old volcano, Little Brother, that originally was about the same size. Middle Sister was the next to form. South Sister is the youngest of the three. The last eruption at South Sister was a mere 1,900 years ago. It seems reasonable that if one of the Three Sisters were to awake from sleep, it would be the youngest. It always seems like the youngest member of any family is the first to rise in the morning. But if South Sister really is waking, it means that the rapidly growing Bend-Redmond area may have a bit more in store than it had planned.

Eastern Washington already knows the consequences of a newly awakened Cascade volcano. In 2005, we celebrated the 25th anniversary of the last time Mount St. Helens woke up. You still can find ash from that eruption between Yakima and Ritzville. Perhaps former Mid-Columbia residents who now live in the Bend-Redmond area may get a chance to relive 1980.

In addition to a nice blanket of ash, Central Oregon residents' next biggest hazard from a possible eruption of South Sister is a mud flow. Geologists call these lahars but anyone that watched scenes from the Toutle River when Mount St. Helens erupted knows what a lahar can do. I still remember houses being carried along by the mud. It reminded me of a river of concrete flowing toward the Columbia River.

According to the U.S. Geological Survey, the biggest mud flow hazards in the Three Sisters area are the McKenzie River, which flows west to the Willamette Valley, and the Upper Deschutes River Valley. The Upper Deschutes River poses more of a hazard to the Sunriver and Bend area than to the residents of the Bend-Redmond area.

The future of the Three Sisters volcanoes and their impact on the Bend-Redmond area remains to be seen. The one thing that geology has taught us is that at least one of those volcanoes will erupt again—sometime. No mere mortal can predict when that next eruption will occur. Perhaps three other sisters, Shakespeare's 'Three Weird Sisters' (the witches) standing around their cauldron in the play MacBeth, foretold the future of the Bend-Redmond area when they said, "Something wicked this way comes."

Index

A

B

C

N

O

P

Q

R

S

T

U

V

W

Y